Food Sovereignty and Uncultivated Biodiversity in South Asia

Food Sovereignty and Uncultivated Biodiversity in South Asia

Essays on the Poverty of Food Policy and the Wealth of the Social Landscape

FARHAD MAZHAR, DANIEL BUCKLES,
P.V. SATHEESH AND FARIDA AKHTER

ACADEMIC FOUNDATION
NEW DELHI

International Development Research Centre
Ottawa • Cairo • Dakar • Montevideo • Nairobi • New Delhi • Singapore

Published in 2007
by
ACADEMIC FOUNDATION
4772-73 / 23 Bharat Ram Road, (23 Ansari Road),
Darya Ganj, New Delhi - 110 002 (India).
Tel : 23245001 / 02 / 03 / 04.
Fax : +91-11-23245005.
E-mail : academic@vsnl.com
www.academicfoundation.com

in association with

INTERNATIONAL DEVELOPMENT RESEARCH CENTRE (IDRC)
PO Box 8500, Ottawa, ON, K1G 3H9
Canada
info@idrc.ca / www.idrc.ca

UBINIG
22/13 Khiljee Road,
Block - B,
Muhammedpur
Dhaka 1207
Bangladesh

DECCAN DEVELOPMENT
SOCIETY (DDS)
Pastapur Village,
Zaheerabad Mandal,
Medak District - 502 220,
Andhra Pradesh, India

Food Sovereignty and Uncultivated Biodiversity in South Asia:
Essays on the Poverty of Food Policy and the Wealth of the Social Landscape
by
Farhad Mazhar, Daniel Buckles,
P.V. Satheesh, and Farida Akhter

(softcover book + DVD)

ISBN-13: 978-81-7188-558-9
ISBN-10: 81-7188-558-6

ISBN (e-book) 978-1-55250-337-8

Printed and bound in India.

Contents

FOREWORD

This book is an eye-opener. It compels us to rethink what constitutes food security, women's knowledge systems, and common pool resources. It makes us see much that we would casually walk past, that we might never taste, that we cannot purchase. Flourishing in the interstices of the cultivated and the uncultivated, the public and the private, the field and the forest, are innumerable leafy greens, fruits, tubers, roots, small fish, grasses, and other forms of food life hidden from our gaze that constitute the daily diet of numerous villagers across South Asia.

Based on a series of studies and practical experiments undertaken over several years by the Deccan Development Society in the drylands of Medak district in Andhra Pradesh (South India) and by UBINIG in the flood-plains of Tangail district in Bangladesh, this slim volume tells us some surprising facts about uncultivated foods. A dietary intake survey by the authors, of 50 families in 10 Bangladesh villages, for instance, found that uncultivated food provided, on average, 65% of the food weight and all of the fodder and fuel needs of very poor landless households, and 34% of the food weight and 20% of the fuel and fodder needs of better-off landed households. In other words, such plants are not just part of the coping strategy that the poor use during seasonal shortages or drought; they are part of the everyday sustenance and key sources of vitamins, minerals and proteins not just of the poor but also of the relatively well-off. In Medak district alone 79 species of uncultivated leafy greens used as food have been documented. Many plants also have medicinal properties. These uncultivated plants living as "partner" plants alongside the cultivated ones inhabit a landscape unobserved by most researchers. They are neither "wild" (unattended) nor "domesticated" (tamed), but something in between community-managed. As the authors emphasize, this uncultivated biodiversity provides not just food security but food sovereignty. It gives people control over a basic need.

Implicit in this volume, however, is also a much-needed warning. The biodiverse ecosystems and the local knowledge systems that sustain these food chains are threatened by the spread of "modern" monocultural, chemically-driven agriculture and by large-scale displacements of people from their familiar environments. Among the undocumented and unmourned hidden costs of such "development" initiatives is precisely the disappearance of knowledge about food systems based on the everyday use of local environments, especially by women. To keep these systems alive is not just the responsibility of the women and their families, nor just of the authors who have long provided these communities their dedicated support it is also the responsibility of all of us as a trust for the future.

Farhad Mazhar, Daniel Buckles, P.V. Satheesh and Farida Akhter, are not simply authors, however. They are activists and environmentalists with a deep respect for the knowledge embedded in the communities they work with and live among; with a strong commitment to empowering these communities, and especially the women; and a firm belief that ecological farming systems can be both efficient and sustainable.

Like the "partner plants", so partnering this book's text are numerous beautiful photographs that say what a text alone cannot—photographs of plants, crops and people, especially of women and children, showing us their relationships with their environs, the seeds they preserve, the foods they gather and cook, the recipes they exchange, indeed their entire way of life.

I recommend this book not only to those interested in food policy and food security, or in agricultural systems and local knowledge systems, but also to those who are simply curious about other communities and food variety. And if you can, you should also visit these villages and savour this feast of diversity and delicacy, as I have done: there is a treat in store here not just for the eyes but also for the palate!

— Bina Agarwal
Professor of Economics,
Institute of Economic Growth, University of Delhi;
and author of *A Field of One's Own: Gender and Land Rights in South Asia*

Introduction

Our Crops – Our Selves

This publication explores the meaning of agriculture. On both the flood plains of Bangladesh and the dry lands of the Deccan Plateau in India, village women cannot imagine the preparation of food for their family members or the raising of livestock without the leafy greens, grasses, roots and tubers, leaves of trees, fruits, small fish, mollusks and other items collected from the borders between fields, along roadsides, in small patches of forest or fallow lands, in ponds and other water bodies as well as in the cultivated fields and homesteads. When women do weeding work or return home from grazing animals they routinely gather these items. They are collected from any space, whether it is their land or not, so long as it can be done in a way that does not interfere with the cultivated crops. From their point of view, the uncultivated plants and animals constitute a prominent place in the local food system and are consequently a vital part of agricultural practice.

The significance of this practice was initially hidden by our own biases based on patriarchal categories of property such as private and common lands, the opposition between homestead and cultivated land, and a narrow focus on the cultivated components of agriculture. Through the research we realized that the very act of collecting uncultivated biodiversity and using it to prepare food or feed livestock brings women into a distinct relationship with cultivated and uncultivated spaces and the gifts of nature. This observation guided us into new territory where food, ecology and culture converge.

...the very act of collecting uncultivated biodiversity and using it to prepare food or feed livestock brings women into a distinct relationship with cultivated and uncultivated spaces and the gifts of nature...

We have worked with disadvantaged social groups for several decades and undertake research to help meet the needs of these communities. Our attention on uncultivated foods should not, however, be seen as a preoccupation solely with an agriculture of the disadvantaged. Rather, it is a lens through which to question mainstream concepts of agriculture, irrespective of the communities with whom we work.

In the context of any agricultural community, documenting all sources of food and fodder is necessary, not only to assess the quantity of food produced per unit of land but also the diversity of food and its availability. A large number of studies show that uncultivated plants and animals are critical to survival during times of extreme hardship.[1] Knowledge of the uses of available plants and animals can make the difference between those that live and those that die. The value of these foods in meeting the more regular nutritional needs of vulnerable groups such as young children is also well documented. Our challenge was to develop an understanding of how to maintain and develop the conditions for ensuring ready access to uncultivated food and fodder sources.

Generating understanding and evidence of the connections between diversity in agriculture, the farming practices that support this diversity and the culture of food systems is a research theme uniting the work of our organizations and broader movements in support of South Asian agriculture such as the South Asia Network on Food, Ecology and Culture (SANFEC).

For two decades the Deccan Development Society (DDS) has studied the diversity of farming in the Medak District of Andhra Pradesh, India some 100 kilometers northwest of Hyderabad, the state capital. The Society evolved through this process in partnership with women's Sanghams, voluntary village level associations of the poor that now make up the core of the organization's governing body. The 5000 women members of the Society represent the poorest of the poor in their village communities and play a defining role in development of the vision and daily practice of the organization. The Society currently has a vision of consolidating the Sanghams into vibrant organs of primary local governance, federated in a strong lobby for women, the poor and Dalits. A host of ongoing dialogues, educational programs, research and land-based activities facilitated by the Society tries to turn this vision into a reality.

Sorting, washing and cutting uncultivated greens is collective work.

Women farmers like Begari Sammamma have a lead role in the Deccan Development Society.

Implementation of the work of the Society, including research, seeks to create the conditions for revival of communities and to readdress gender-based limitations on access to and control over the resources of farming and rural livelihoods. The research on uncultivated food consequently took advantage of the many years of DDS experience with participatory methods such as resource mapping and matrix ranking developed by village women themselves and men from their communities. These methods are part and parcel of the mode of research and exploration of problems, people and options used regularly by women in the Sanghams, along with other methods of documentation and dialogue such as festivals celebrating the biological and cultural diversity of the Deccan.

The research in Bangladesh was facilitated by UBINIG, a research and advocacy organization. The term UBINIG is a Bengali acronym standing for Policy Research for Development Alternative, and reflects its origins in the early 1980s as a study circle of university students and development activists in Dhaka. Village-based programs were initiated in 1986 with weavers in the District of Tangail some 100 kilometers northwest of the capital. A handloom-weaving initiative succeeded in organizing craft weavers and enhancing the efficiency of textile production processes and marketing, thereby arresting the rapid loss of an historically significant rural industry.

In 1988 a severe flood in Tangail affecting both weaving and farming families brought UBINIG into closer contact with farmers in the villages. It soon became apparent that farmers were facing a broader crisis than flooding. Women in particular complained bitterly about the impact of pesticides and herbicides on their health and the health of their children as well as the loss of the leafy greens in agricultural fields and fish in the various water bodies. Moreover, pesticides were seen as a killer of human beings, as they were used by bankrupt and despondent farmers to commit suicide and by various people to bring on miscarriages and even murder unwanted wives and girl children.

Farmer women took the lead in calling for an end to the use

Video amplifies the voices of non-literate women like Chinna Narsamma, a farmer film-maker in India.

of pesticides in their communities, and the development of alternative ways of farming. Over time, groups of farmers experimented with green manures, composting and mixed cropping, eventually coming to the conclusion that they did not need to depend on pesticides.

Ten "rules" of ecological agriculture emerged, providing the basis for development of the "Nayakrishi Andolon" (New Agriculture Movement), a community-based farmer movement now numbering more than 170,000 farm families in 15 different districts throughout Bangladesh. Several local governments have declared their territories pesticide-free zones, joining forces with Nayakrishi farmers to stop the sale and use of pesticide within their boundaries.

Throughout this process UBINIG played a catalytic role for farmers and served as a source of information on the farming practices of other successful communities and new options coming from the scientific community. The collaboration between UBINIG and the Nayakrishi Andolon involves a constant dialogue on practical ways for farmers to reconstruct the material foundations of communities capable of achieving economic prosperity and social justice. The science of popular wisdom generated through daily experiments and the practice of farming provides the ground from which to critique the present overbearing character of 'science' while at the same time avoiding a romantic view of current farmer knowledge.

The kind of agriculture and discourse that has emerged from this interaction, while seen by outsiders as conservationist and critical of globalization, is understood by the Nayakrishi farmers as cultivating "ananda", which literally means joy or happiness. Planting the seeds of joy as understood in the local culture means creating a farming practice that rests on joyful or happy relationships both within and between human communities and external nature. It is essentially a lifestyle that is not only a very advanced way of farming but also inspired by the unity of the universal human spirit. From this perspective, integrating the uncultivated plants and animals in the Nayakrishi practice of farming, whether they serve an immediate use to people or not, is a pragmatic and necessary step.

This publication is a genuine attempt to share what we have learned from farming communities during a series of research studies and practical experiments conducted over several years in Bangladesh and India. The combination in the essays of facts from the research and passionate argument reflects our desire to inspire like-minded organizations to defend the biodiversity and sovereignty of South Asian agricultural communities.

The publication also presents new policy directions. Discussion of policy issues and front-line debates on food security and poverty alleviation is an attempt to engage with sympathetic but skeptical policy makers in need of new directions. Even the most supportive among this class of democratic representatives are currently stuck in a quagmire of flawed techno-economic arguments with very little genuine relationship to the ground realities of people, their issues, problems and coping mechanisms. Among these policy makers we hope to find readers that will be

Seed saving is a right affirmed by women farmers.

The Water Lily (Nelumbo nucifera) has edible rhizomes (root nodules).

jolted out of their present conceptions to seriously rethink the meaning of agriculture and food policy parameters. The publication seeks to provide a new set of parameters they can pursue and, more immediately, to bring confidence in people's agriculture into their policy thinking.

The publication raises as well research questions and conceptual concerns posed by different academic disciplines including economics, ecology, anthropology, political science and women's studies. It tries to avoid the tendency to compartmentalize disciplines and methodologies by pointing out the economic role of food in cultural and biological systems while at the same time drawing attention to the cultural and ecological dimensions of food in economic relations. It also tries to bridge the gap between class and gender perspectives by illuminating feminist strategies in people's designs for livelihood. Our intention is primarily to convince researchers that uncultivated foods are not simply a food of the poor but rather a significant concept for critically rethinking notions of space, food systems and gender relations in agriculture. The evidence and arguments are developed more with an activist's eagerness than an academician's depth, but will hopefully point in new ways to further research.

Finally, we have designed the publication in a way that supports sharing of information among farmers grounded in oral cultures. The general picture we hope arises in the mind of the reader, and which informs our own perspective, is that agriculture is determined by the diversity and dynamism of local food systems embedded in specific cultures and ecosystems. Developing food policies, farming practices and community relations compatible with this understanding of agriculture is, in our view, critical not only to the conservation of the uncultivated biodiversity of South Asia but also to the further development of agriculture as a life-affirming practice and deeply human undertaking.

Chaina with her daughter and neighbour Anwara.

[1] For reviews of evidence regarding the role of "wild" foods under conditions of hardship and more generally in contemporary agricultural systems, see;

Grivetti, L. E. and B. M. Ogle 2000. Value of Traditional Foods in Meeting Macro- and Micronutrient Needs: The Wild Plant Connection. Nutrition Research Reviews. 13: 1-16;

Guijit, I., Hinchcliffe, F. and Melnyk, M 1995. The Hidden Harvest: The Value of Wild Resources in Agricultural Systems: A Summary. IIED, London;

Kuhnlein, H. and Pelt, G. H. 1997. Culture, Environment, and Food to Prevent Vitamin A Deficiency. INFDC, US and IDRC, Canada;

Prescott-Allen, R & C. 1999. Assessing the Sustainability of Uses of Wild Species - Case Studies and Initial Assessment Procedure. Occasional Paper of the IUCN Species Survival Commission No. 12. IUCN Publications Services Unit. Cambridge, UK;

Scoones I., Melynk M, Pretty J.N. 1992. The Hidden Harvest, Wild Foods and Agricultural Systems. A Literature Review and Annotated Bibliography, IIED, London.

A matrix asessing the contribution of different grains to farmer livelihoods.

What is Agriculture?

Diversity on the Flood Plains of Bangladesh

The people of Bangladesh are known as the "Mache-Bhate Bangalee" (the fish and rice eating Bengali), a nickname particularly fitting for communities living on a flood plain. Our study of uncultivated biodiversity in Bangladesh draws mainly from communities in the District of Tangail, some 90 kms north-west of Dhaka. The District is in the central part of the Ganges-Brahmaputra-Meghna delta, an enormous flood plain ecosystem. Throughout the District old river ways have left their traces in moribund channels, now forming shallow, swampy areas where cultivation of rice is practiced. During the rainy season, the areas adjoining these water bodies are flooded, providing spawning grounds for numerous fish species.[1]

Historically, Tangail was known for its large variety of freshwater fish species and its deepwater rice varieties. About 75% of the population is directly dependent on agriculture for their livelihood, although weaving is also an important activity in some villages.[2] Most farmers work 1-3 acres of land but 25% of the population is landless, having only homestead land. These people work as agricultural laborers as well as in a variety of occupations such as rickshaw pulling, craft work, petty trade, etc.

The area has a monsoon type climate typical of the Indian subcontinent and total annual rainfall of more than 1700 mm on average. Crops are grown throughout the year in three distinct seasons. The pre-monsoon season (Aus) runs from the end of March to May and is characterized by periodic heavy thunder storms and occasional hail-storms and cyclones. The hotter monsoon season, known as Aman, is from May to September, during which some 80% of the total annual rainfall occurs. The winter season (Boro), from mid-October to early March, is relatively dry.

Rice is the most common crop in both Tangail and Bangladesh as a whole. The largest area of rice is planted during the monsoon, by transplanting in deepwater. Pre-monsoon rice is also common, and is the first rice harvest of the year. Winter rice is sown on some farms but other crops are more common in the winter season, including a wide range of vegetables, pulses and oil seeds. A typical farm grows both rice and vegetable crops in each season. The diversity of cultivated crops is very high, with some 55 plant species and more than 453 varieties documented in a single village.

The number of uncultivated species used as food is also very large. Some 102 species of leafy greens associated with agricultural fields, homesteads and common areas were identified by local people as food, mainly herbs, creepers, aquatic plants, shrubs and trees.[3] They are most abundant during the monsoon, when other food plants are not ready for harvest, but are also common in the winter and pre-monsoon periods.

Uncultivated plants in and around agricultural fields and homesteads are considered "partner plants" in the sense that they grow in association with the cultivated plants. For

A rich habitat for uncultivated fish and edible water plants is lost when beels (large depressions) are damned, ploughed and seeded in a monoculture.

example, Bathua grows on its own in the wheat or mixed crop mustard field and can be collected when cleaning the agricultural field for the main crop or at any other time. The plants found in the common areas such as along roadsides, boundaries between fields, water bodies, in forest patches, in fallow pastures and around water sources are usually different species, although some grow in several different habitats. For example, Dondokolosh is a small plant that grows both in fields and in forests and is used as food and medicine. The taste, nutritional and medicinal qualities of the same plant are known to be different when collected in agricultural fields compared to uncultivated lands, a result of the dynamic interaction of the soil conditions and the plant genotype.

Uncultivated plants are typically plucked rather than uprooted so that they will regenerate and provide food again. They are in this sense similar to another category of uncultivated foods recognized by local people, that is, cultivated plants that provide edible leaves and shoots in addition to the primary product of cultivation. This includes species such as mustard which is cultivated for the oil seed and jute which is cultivated for its fibrous stalk. The leaves of both plants are also plucked for food. In fact, plucking the leaves in the early part of the growing season is considered by farmers to be good for the plant and is therefore allowed

Shenchi sak and many other greens are an important part of the food system.

when done with care. These "derived foods" include as well the leaves of pumpkins, sweet gourd, white gourd, and other vegetables grown in the homestead and plucked for food in small quantities almost everyday before the fruit is formed. Flowers of some cultivated plants that are unlikely to be fertilized are collected as well.

Uncultivated plants are importants foods for animals.

Some 58% of the households in the communities of Tangail have cows or bullocks, usually one or two animals. Almost all households have small ruminants including goats, chickens and ducks. For example, Rabeya of Nallapara village has one cow, one goat, 13 hens, and 23 ducks and ducklings. This is typical not only of farming households but also many landless farmers with homesteads.

There is also a large variety of fish in the common water bodies, paddy fields, ponds, canals and rivers. In Hinganagar village alone some 69 different varieties of fish were recently documented. Many of these are highly nutritious small fish collected by children in the course of their day, on the way home from school, while bathing or simply playing with their friends.

[1] *Bangladesh has more inland fish species than all of Europe, with estimates numbering from 260 to over 500 distinct species when the river-dependent fishes of the Bay of Bengal are also considered. The inland fishery is one of the most productive and biologically rich in the world, and a major source of food for the population. See Minkin, S. M Mokhlesur Rahman and S. Halder. 1997. "Fish Biodiversity, Human Nutrition and Environmental Restoration in Bangladesh", In Chu-fa Tsai and M. Youssouf Ali (Eds.) Open Water Fisheries of Bangladesh. University Press Limited, Dhaka, Bangladesh.*

[2] *Our study draws mainly from 90 villages in several Unions of the district, with a total population of about 820,000 people.*

[3] *UBINIG. 2002. Our Uncultivated Leafy Vegetables. Narigrantha Prabartana, Dhaka, Bangladesh.*

Diversity on the Dry Lands of the Deccan Plateau

The dry lands of the Deccan Plateau in India present a sharply contrasting environment compared to Bangladesh yet they also host a vast array of cultivated and uncultivated biodiversity used by people as food or fodder. The Deccan

Women ARE agriculture.

Plateau of South Central India hosts some of the poorest populations in India and is one of the most difficult of farming environments. The region receives an annual average rainfall of less than 900 mm, with mean daily highs of 40° C in May, the hottest month of the year. The natural vegetation is very sparse and the soils mainly poor, red laterites. Large tracts of land are degraded pastures owned by the government or absentee landlords.

The publication draws on information from about 75 villages in the Medak District Andhra Pradesh in the Deccan, villages with a total population of some 100,000 people.[1] Most are small farmers with 1-3 acres of land gifted during the feudal period by their landlords or in the more recent era of government-led land reforms. The better-off farmers, comprising about 20% of the population, have farms of more than 3 acres, with a large proportion of their land comprised of better black soils and wells for irrigation. Some 17% of the population are landless Dalits, a socially excluded and vulnerable group occupying the bottom of the Indian caste hierarchy. A few households (about 5% of the population) are engaged in occupations related to farming such as carpentry, pottery and petty trade. A small number of people (about 3%) are members of tribal groups and subaltern castes such as grave diggers, flower gatherers or washers of clothes.

The land supports a wide variety of agricultural crops including sorghum, a range of millets, pulses and oilseeds, all of which grow under rain-fed conditions. The Kharif or monsoon crops are planted in June with the first heavy rains and harvested from August through to February. Rabi or winter season crops are planted in October in the deeper red soils and in the few places with alluvial black soils and are harvested by March. The winter crops include wheat, dry-sown rice, linseed, chickpea, field pea, lentils, lathyrus, coriander, and groundnut. The local people call some of these post monsoon crops "Satyam Pantalu" or "Crops of Truth", a metaphor for crops that are hardy enough to grow on the available soil moisture and limited soil fertility.

A typical rain-fed farm in the area grows between 8 and 10 crops in each season. Within the species diversity of the cultivated crops there is also a large varietal diversity. For example, during a recent monsoon season Masangari Chinna Narsamma Adivappa grew on her 2.5 acres of red soils three varieties of sorghum and two varieties of pigeonpea along with greengram, redgram, finger millet, cowpea, Dollichos lablab and hibiscus. She is not atypical in her region. Surrounding her farm are 12 or so other farms of a similar size and condition, except for one farm where the farmer has dug an open well and irrigates his sugarcane, turmeric and chilies during the winter season.

A few farmers in the study area, making up about 6% of the farming population, grow an even larger number of species and varieties. On a high diversity farm like that of Gangwar Manemma, the following monsoon crops grow in a single year on a three acre expanse: four varieties of sorghum, three varieties of pigeon pea, three varieties of cowpea, three varieties of foxtail millet, little millet, finger millet, two varieties of hibiscus, Dollichos lablab, yellow cucumber, groundnut, four varieties of redgram, three varieties of field beans, niger and horsegram.

Within this diverse cropping system, the number of species of uncultivated food plants used by people greatly exceeds the number of cultivated species. They are most abundant during the monsoon when the heavy showers coax the land to exhibit all her fertility. The land responds with a large variety of herbaceous greens, creepers and new growth on trees and shrubs. Many species that germinate during the later stages of the monsoon are also abundant and readily available during the winter season as there is enough residual soil moisture to produce mature plants even in the dry, red soils. The heavier and more fertile black soils also support leafy greens during the winter season. Trees, but also some creepers, have edible parts year round. A recent study of this diversity identified 79 species of uncultivated leafy greens used as food along with many other species of uncultivated plants including roots, tubers and fruits.[2] As in Bangladesh, there are also many cultivated plants with edible parts secondary to their main use. For example, Amaranthus is cultivated for its grain, but its leaves are plucked for food throughout the growing season without damaging the growing plant.

In addition to the plant diversity, livestock diversity also forms an important part of the agricultural system of Medak. Cattle, buffalos, sheep and goats are common, as are chickens. A few non-domesticated animal species such as monitor lizards, wild pigs, rabbits and fish are also recognized and used as food by people in the region. Together the cultivated and uncultivated plants and animals provide for a wide variety of foods that are part of the farmers' diet and culture.

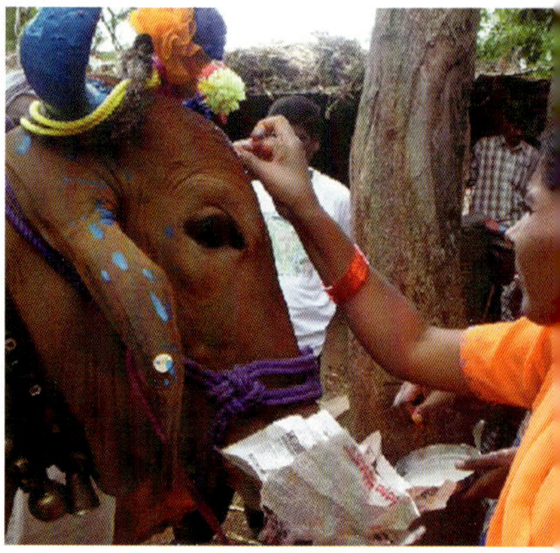

[1] *Handbook of Mandal Statistics, Medak District, Government of India.*

[2] *The first comprehensive list of uncultivated plants in the area was compiled by S. Reddy and S. Yesudas in 2000. See "A Participatory Rural Appraisal on Foods" by Eedulapalle Bayavva, Hoti B Sangamma, Basantpur Santhoshamma, Raikode Balamma, Pastapur Pushpalamma, Metlakunta Paramma, Humnapur Kashamma, Pyalaram Thukkamma, Kalmela Kamalamma Narsapur Bagamma and Kalmela Narsamma. Deccan Development Society, Pastapur, India.*

The Politics of Weeds - and the Decline of Farming Systems

"What do you mean by weeds? There is nothing like a weed in our agriculture. We eat whatever grows on our land. If we can't, our cattle eat them. There is no weed on our land. .... Yes. I have seen weeds in those lands where they use Sarkari Eruvu (chemical fertilizers). In fact weeds started with Sarkari Eruvu. You know why? All that you call weeds are born out of tiny seeds on our lands. They are all tender and edible. Because we don't apply Sarkari Eruvu on our lands our soils are soft. They allow the tiny seeds to germinate. But on those lands where Sarkari Eruvu is applied, the land gets encrusted. The top layers get hard. And no tiny seed can germinate through those soils. What plants do germinate and come out of those soils are hard and inedible. You see, Sarkari Eruvu really gave birth to weeds. Not our agriculture. In ours, all plants are edible."

Eeramma, 62, a woman farmer from village Shekhapur, Medak District, Deccan, India

What Eeramma says is a profound truth which has tragically escaped the attention of agricultural research and policy. In Eeramma's kind of agriculture, which is the hallmark of all traditional farming in South Asia, the entire social landscape is a host to plants, both cultivated and uncultivated. And all of them are edible. They are food for humans or animals. What are "weeds" to the world of agricultural science are welcomed plants from Eeramma's perspective.

The sharp dichotomy of crops and weeds, while a commonplace distinction, is part of a broader discourse on the development of monocultures, propriety claims to seed, and corporate control of the global food system. Modern agriculture focuses primarily on maximizing the grain yield of a few plants species required by industrial food processors, including rice, wheat, maize, beans and a handful of other species. The contemporary science of plant breeding evolved to support this process by focussing on improving the design of a small number of varieties of these privileged species. The diversity of plant genetic resources in agriculture and the interaction of genotypes with diverse social and biophysical environments was marginalized to create the technology of monocultures.

The emphasis in modern agriculture on monocultures goes hand in hand with the chemicalization of agriculture. Wherever agricultural chemicals are used, the seed and cropping system alters. For example, the normal sorghum-pigeon pea-cowpea mixed cropping system of the Deccan Plateau cannot tolerate herbicide use: herbicides applied when one species needs weeding negatively affect the growth of the other intercropped plants. The use of herbicides inevitably requires a shift to a monocrop such as cotton or potato that can pay for the chemicals purchased. It also destroys the leafy greens in the field and renders inedible the plants along field edges and pathways.

Farmers, not registered trade marks, feed the world.

The effects of changing cropping practices and the chemicalization of agriculture are apparent if we examine what happens to livestock in a system based on monocultures. On the one hand, herbicide use makes it impossible to get any fodder from uncultivated plants in the field. On the other hand, the monocrop itself, whether it is potato or cotton, produces no residues which can be used as fodder.

Compare this with the mixed cropping strategy of the Deccan plateau. These fields provide all that the cattle need: a continuous supply of green fodder during the cropping cycle (the so-called "weeds"), dry fodder from the crop residues (sorghum and pearl millet stalk, little millet straw, etc.), feed from the husk of the cultivated grains and pulses (paddy, pigeon pea, *mung dal*, etc.), and a host of creepers which are central to the farming system (cowpea, dolichos lablab, etc.)

The mixed cropping strategy and related crop rotations provide space and time for these "partner plants" to develop between the cultivated plants and the practice of hand weeding allows for the selective management and harvest of these food and fodder sources. Two acres of mixed crop combinations in the Deccan can support three head of cattle in an average year. In return, cattle supply almost all the

Video clips from "why are Warangal farmers angry with BT Cotton?"

nutrition that the system needs in the form of urine spilled on the soil and dung laid on the land by farmers before ploughing. Produce comes out of the field, nutrients go into the field.

The cattle also supply all the draught power for transportation of goods to and from the field. The type of plow used in Medak turns only the surface layers of the soil, making it possible for the tiny seeds of a host of uncultivated plants to germinate easily. These seeds are carried into the fields from surrounding areas by the wind or in the animal manure and compost, providing a continuous supply of diversity in the agricultural system. Goats, chickens, ducks and other animals are also supported by the combined production of cultivated and uncultivated diversity, both in agricultural fields and in surrounding common areas. This is a powerful system of energy recycling and integration across agricultural activities which monocultures cannot create.

Conceptually, the notion of a single, improved species is flawed because all other plants, and even other varieties of a species, are implicitly reclassified as "weeds", to be excluded from the space of the priviledged plant. It justifies and even requires destructive, propriety technologies, seen most potently in the business strategy of the Monsanto Corporation. Seeds bred by the company to be tolerant to the herbicide Round Up® are essentially vehicles for the sale of the herbicide, a trade that accounts for most of the company's profits. The sterilization of seeds before they even leave the corporate factory, a plan provisionally blocked by global resistance to the "terminator" technology, would cement control over agricultural seed, the starting point in the entire global food system.[1]

The single plant paradigm gives rise as well to the misleading notion that global food systems are or should become based on the production, circulation and consumption of a few "staple" crops, distributed around the world through a global trading system. While global statistics track and suggest that some 15 to 30 species feed the world, more precise and regionally sensitive reviews indicate that the contribution of other plant species to local and regional food systems has been greatly underestimated. For example, Christine and Robert Prescott-Allen identify 103 species that seem to provide the largest part of the world's plant food supply by volume.[2] They recognize that even this estimate does not take into account the role of food diversity in meeting the regular nutritional requirements of different populations and many undocumented regional and local variations.

More generally, the corporate push for food and agricultural systems built on fewer and fewer plant species ignores the

historical development of food technology in different cultures and diversity as an integral part of the cuisine system. In the context of South Asia, food cannot be separated from the position plants and animals occupy in the culinary practices of the local cultures. Bethua *shak* (Chenopodium album L. of the family Chinopodiaceae) is not a cultivated plant in Bengal, but it's hard to imagine the rural cuisine of Bengal without it.[3] It is an important leafy vegetable just like any cultivated cabbage or spinach. Consider the

Bengali literary epic Monosha Mongol. In this story Sanaka, the wife of Chand Sawdagar, cooks 10 leafy vegetables, including the leaves of bethua *shak*, chalta, gima *shak*, and kumra *shak*. All are uncultivated, celebrated as delicacies and the supreme expression of her art of cuisine. Strikingly, bethua is classified as a "weed" by Bangladeshi scientists.[4]

The perspective of farmers on the relationship between the cultivated and the uncultivated is more nuanced, recognizing the ecological, cultural and ethical determinations of food production as an economic activity. While the term agacha is used by Bangladeshi farmers to refer to plants that are not intended for cultivation, it does not imply that the plants are useless and absolutely unwanted. Under most conditions the uncultivated plant species in agricultural fields play a positive role in the recycling and conservation of nutrients and soil moisture, as traps for some insects that harm the main crop and habitat for others that help. In terms of agricultural practice, these plants are managed, not destroyed, to reduce competition with the cultivated plants while at the same time retaining their potential for regeneration as food plants, for humans or animals.

From this perspective, the notion of "weed" has no technical validity. There are always a multitude of specific individual and community needs different plants fulfil. For example, the grasses listed in Table One grow in rice fields in Bangladesh where they are an important source of fodder in a context with almost no grazing land. Fodder for animals is sourced from the same land that is producing food for human consumption. What are seen by agricultural scientists as "weeds" actually make up part of the harvest from the land.

The human relationship with plants is not static or one-dimensional. During periods of hardship knowing how to use particular plants and animals is critical to survival. Ensuring that they are available in the environment consequently takes on a new importance in more regular circumstances. The continuity of diversity in an agricultural environment also supports multiple human experiments with plants and animals in a dynamic and meaningful relationship. These experiments may be designed to meet the functional needs of the community or they may be undertaken out of intellectual curiosity, as an inspiration for spiritual and cultural experience, or for ethical reasons. Many South Asian farming communities believe that taking care of plant and animal life forms is a way to seek the meaning of human existence or communion with God.

From these perspectives, each and every being is part of a living reality, with a place in the order of the world that constitutes the community. Keeping this place secured over time is the first condition by which human communities ensure the conservation of life, whether it satisfies an immediate human need or not. 'Apanzala', a term used in Bangladesh for uncultivated plants, evokes these very deep ecological and spiritual associations. It refers to a life that has appeared on its own before the community. It implies limits to ownership of the plants, the natural rights of others to have access to the plants, and certain obligations to care for them.

The practice of agriculture in South Asia continuously blurs the distinction between the cultivated and the uncultivated, the crop and the weed, not only in times of hardship but in the everyday process of farming, feeding livestock, cooking and community relations. From a farmers' perspective there is no clear divide. Uncultivated plants are not "wild", as though they were left unattended or without any implicit or explicit community management. Nor are they "domesticated" in the sense of being completely tamed for human use. Rather, they belong to the social landscape, not simply as a material entity, but also as a cultural entity, from which communities draw their food and construct social relations. The horizon of the social landscape is like a spectrum or a continuum. At any one spot there are always surprises of nature: a plant that had no direct use yesterday suddenly becomes part of cuisine, an element of a concoction for medicine, a spice, or what not. In another location there are plants ready to be used as food, if necessary, during famine, flood, or other conditions of stress. The cultivated plants indeed occupy the narrowest place in the diverse richness of this spectrum. Most importantly, there is no meaning in the unwarranted contradiction, antagonism, or dichotomy between cultivated and uncultivated plants, because both are equally crucial, important, and useful. The ingenuity of a community lies in continuously redrawing the margins in the context of their day-to-day struggle for survival.

Apanzala refers to a life that has limits to ownership of the plants, and certain obligations to care

Table One: Partner Plants in a Tangail Rice Field

Bangla name (local name)	Scientific name	Seasonality and habitat
Uri (Jhora Dhan)	Saceialepis indica	Available in the Aus and Aman rice fields.
Dudhalu (Dudhnal)	Eragrastis elianensis (All) vingolo	Available in the higher lands where Aus rice is cultivated
Chto Dhalu (Choto Dudhnol)	Eragrastis asperaness	Available in the mustard & wheat field and in fields where pulses are grown.
Hiyal Lengta (Sabuj Shial Leja)	Setaria viridis (L)	Available in the Aus and Aman rice field and among winter crops
Hiyal Hengta (Holud Shial Leja)	Setaria glanuea (L)	Available in the Aus and Aman rice field and among winter crops
Durba (Jol Durba)	Sporobolus tremulus	Available in the Aus and Aman rice field
Kharma (Shakta Khagra)	Cyperas pilosus	Available in Aus and Aman rice field
Painnya Chopa (Pani Sholango)	Cyperas distans	Available in Aus and Aman rice field
Malancha & Helencha (Helencha)	Enhydra Fluetuas lour	Available in Aus and Aman rice field and in aisles.
Kanai Ghash-1 (Knaibashi)	Commilina bengha lensisl	Available in Aus and Aman rice field, in aisles and along the road side.
Kanai Ghash-2 (Kanainala)	Cyanotis axillaris	Grows in Aus and Aman rice and in aisles
Shenchi (Chanchi)	Alternan thera Sessilis	Grows in Aus and Aman rice and in aisles and also in the road side

appeared on its own before the community. It implies the the natural rights of others to have access to the plants, for them.

[1] *The ETC Group (formerly known as RAFI) monitors the ownership and control of technologies and the consolidation of corporate power, especially in the field of agriculture (http://etcgroup.org). GRAIN also monitors developments in the seed industry that have a potential impact on the use of agricultural biodiversity based on people's control over genetic resources and local knowledge (http://www.grain.org).*

[2] *Christine and Robert Prescott-Allen 1990. How Many Plants Feed the World? Conservation Biology, Vol. 4, No.4, pp 365-374. December 1990; Prescott-Allen, R & C. 1999. Assessing the Sustainability of Uses of Wild Species - Case Studies and Initial Assessment Procedure. Occasional Paper of the IUCN Species Survival Commission No. 12. IUCN Publications Services Unit. Cambridge, UK. A comprehensive assessment of global biodiversity, including agricultural crops, is presented in Heywood, V.H. 1995. The Global Biodiversity Assessment. United Nations Environment Programme. Cambridge University Press, Cambridge.*

[3] *In English Bethua is known as lamb's-quarters, fathen, dog's tooth grass, and goosefoot. Shak is the Bengali term for leafy green.*

[4] *The book "Weeds of Bangladesh" (Karim and Hasan Kabir, 1995) published by the Bangla Academy in Dhaka, Bangladesh lists bethua as a weed in agricultural fields, without even mentioning its role in Bengali culture and cuisine.*

[5] *Information collected in Atia, Delduar, Kanchanpur and Gobindadashi of Tangail District.*

Why do Dry Land Farmers in the Deccan Practice Ecological Agriculture?

This essay is based on an oral presentation by Ms. Bidakanne Chandramma and Mr. Yerrolla Jayappa, two farmers of the Deccan Plateau.

Ecological farming in our region of the Deccan is based on diversity. The diversity is directly related to the variety of foods that are a part of our diet and food culture. All of our crops are nutritionally rich. The host of uncultivated greens that we can harvest from our fields as a result of the practice of ecological agriculture also helps meet the nutritional needs of our families. They are either food for humans, cattle or medicine plants for both.

The diverse crops also help farmers preserve soil fertility and ensure against various climatic and pest problems. The diversity includes early maturing varieties and late maturing varieties so as to correspond to the early and late monsoons. It includes varieties that can grow with little rainfall, varieties that can manage with excess rainfall and varieties that thrive with normal rainfall. In this manner we protect our farms against the vagaries of weather.

The medicinal properties of different foods also enters into our thinking. Each food has a special medicinal quality that is important. Little millet is eaten for its cooling quality and foxtail millet for its warmth. Horsegram clears the body of kidney stones while sorghum is good for the diabetics. Almost every crop combines this food-is-medicine property and is consequently a very important part of our healthcare.

The diverse needs of our culture and rituals also prompts us to practice ecological agriculture. Each of the festivals of the region demands a particular food crop. Together, the scores of festivals celebrated throughout the region means that there is a religious purpose for diversity in agriculture.

There is no seed on our farms that does not answer the need of medicine or fodder or food or nutrition or ritual. Every crop is good for the soil, good for human beings and good for the animals. It is this relationship that sustains agriculture.

ayappa and Chandramma preparing a village map.

Every crop is good for the soil, good for human beings and good for the animals. It is this relationship that sustains agriculture.

Penta Pooja – Worship of the Manure Heap

April is the month for Penta Pooja, worship of the manure heap. Farmyard manure is a catalyst for the soil and one of the most treasured parts of farming in the Deccan, comparable only to seeds. To begin the farming season the woman of the house lights a lamp, breaks a coconut and sprinkles vermilion and turmeric, the two most reverential tools of worship, before the manure heap. She does a aarti, turning a lighted lamp in a circular motion in front of the manure heap, exactly as she would worship the most important gods in the Hindu pantheon.

Eruvaa! Adokka Laxmi [Manure! That is the goddess of wealth] is a popular saying among the Deccan farmers. But most of the time this is interpreted in scientific circles as a sentimental statement of ignorant farmers. A survey of 79 households in the village of Pastapur in the Medak District shines a different light on this supposedly ignorant statement. A random sample was drawn representing 20% of the village, stratified by the categories of larger farmers, small farmers and the landless. The findings show that the large farmers used an estimated 5,500 tons of farmyard manure on their fields, of which only about 2,425 tons came from their own manure heaps. The remaining 3,075 tons was purchased from other villagers who had either no land and therefore no use for the manure they had generated, or had excess manure that they could not use on their own lands. The total value of the manure used or purchased by the larger farmers was an estimated INR 425,000. Rental of cart time totaling an estimated INR 125,000 and around 8,000 person days of paid wages amounting to an estimated INR 240,000 was also employed to spread the manure on the fields.

The smaller farmers and landless households produced an additional 3,895 tons of farmyard manure, with a monetary value of INR 310,000. The tasks of applying this manure to the fields of the smaller farmers generated demand for about 5,250 person days of wage labor, with a monetary value of an estimated INR 160,000.

For the village as a whole, the financial and in-kind transactions involving farmyard manure amounted to an estimated INR 1,260,000, equivalent to some $30,000 USD or 84% of the entire annual budget of the Pastapur Village Council. It is no wonder that farmers exclaim: Eruvaa! Adokka Laxmi.

Biodiversity and the Technology of Cooking

The collection of uncultivated plants is only part of what farming people do to turn them into food. Collection merges seamlessly with the science and technology of cooking. While uncultivated plants are available in various spaces and at various times of the year, the cook must know what she wants to collect and why. This knowledge determines the collection, and alters the dividing line between the edible and the non-edible.

To the rural women of Bangladesh 'shak' are not simply leafy greens but rather any uncultivated plant available in and around the household that can be rendered edible by the technology of cooking. They are 'shak' if one knows how to cook them. The link between plant genetic resources and the local cuisine system is direct, reflecting not only a particular world view or epistemology of food but also the day to day practice of local science and technology.

The relationship between the *shak*, the surrounding environment and the cooking fire is also direct and very

immediate, revealing a process of food production, circulation and consumption distinct from that of other food stuffs such as 'shabji', the Bengali word for vegetable. *Shak*, unlike *shabji*, are not normally purchased in the market. Typically, they are collected while the cook prepares to light the 'chula', a conical clay stand used to concentrate the fire on the cooking pot. Girls are sent with a bamboo basket or tray to collect leafy greens indicated by the cook, whether it be a mother or grandmother. The boys are also helpful in the collection of leafy greens, roots and tubers from the water bodies during the rainy season and typically combine this activity with the collection of small fish using a small

aluminum pot and small fishing net. Within 15 to 30 minutes the children gather the leaves and bring them home where they are sorted, checked for spider eggs and other insects and washed until the water is clean of dirt. They are then drained on a bamboo tray or kept in a bowl, depending on whether the water on the leaves is to be used during cooking. They are cooked as soon as they are collected, unless they are left overnight on the roof as a part of the cooking process for the next day's morning meal. When they do start appearing in the market, as some have in recent years, they lose the grace of being called *shak* and take on a new meaning as *shabji* – a vegetable mediated by the market.

The collection and analysis of recipes and the cooking styles of women in Tangail shed light on the relationship of uncultivated plants to fire and the cooking pot, as well as other cooking parameters. *Bhaji* is a common preparation, made from chopped leafy greens, onion and garlic fried in mustard oil. A bamboo stick is used to stir the ingredients. Red or green chilies are added, to taste, alternating stirring to distribute the heat and covering with a lid. The determining parameter in the definition of *Bhaji* is the absence of turmeric, a spice used in Bangladesh to conserve cooked food. Turmeric is an organic biocide, a useful spice in a tropical and humid environment as it slows down the fermentation process and the growth of harmful bacteria. *Bhaji*, by virtue of the absence of tumeric, refers to a preparation that is consumed immediately after cooking.

A *Bhorta* is a mashed preparation of leafy greens, mixed with onion, garlic, green chilies and (optionally) mustard oil. The process of preparation is derived from the cooking of other dishes rather than the *Bhorta* itself. The ingredients may be steamed lightly along with rice or simply mashed by hand. A *Bhorta* is only partially cooked at most and the final preparation does not contain added water. In some areas of Bangladesh *Bhorta* is also known as *Chana* or *Bata*. Because of this particular cooking process the part of the plant collected for a *Bhorta* must be tender and very young. Not all leafy greens are good for *Bhorta*.

Charchari is a dry dish of greens cooked over a very hot fire with mustard oil, small fish and spices such as coriander or chili paste. The dish is ready when the combination makes a distinctive 'char char' sound indicating that the water has dried up and the greens, fish and oil are properly mixed. *Charchari* is a delicacy in the Bengali cuisine system similar to the 'Lahore Karai' – a particular way of cooking meat in Lahore, Pakistan. The fish species most suitable for *Charchari* are collected from open water (as opposed to ponds) and are susceptible to rapid decomposition. Women know to use the greens *Pipul Shak* (Piper longum) and *Gonail Shak* (Paederia foetida) in this dish as they go very well with partially fermented or rotten fish. Plants that are cooked with rotten fish have a very significant role in rural diets and are an important source of nutrition.

Bhaji (cooked greens) and Water Lilly nodules prepared by Rabeya Begum.

Ghonto is a dish of cultivated plants such as potatoes and eggplant mixed with uncultivated shak. It cannot be made up of cultivated or uncultivated plants alone as it is by definition the mixing of unrelated plants or unrelated parts of plants in the medium of water. The use of spices in *Ghonto* is minimal, focusing instead on using water to extract and make available nutritional and medicinal properties. The mixing of unrelated items is needed to manage tastes in the dish as many valuable plants or parts of plants are very bitter or tasteless when cooked alone. *Ghonto* is consequently a kind of cooking method used for plants that are otherwise not so edible but can add to the culinary repetoire.

A dish may be cooked as a single green or as a mixture in very precise combinations of species and in association with particular cooking styles. These change with the seasons. For example, *Kata Notey shak* (Amaranthus spinosus) is available all round the year, even in the summer, but is not consumed until the winter because that is when it is tastier. *Gima shak* (Hydroctyle rotundifolia) is eaten during the dry month of April even though it is relatively scarce at that time because in the dry season it has a particular medicinal value.

Kolmi shak (Ipomoea aquatica), a green collected from water bodies, is available year round but is prepared differently according to the season. When eaten in the rainy season it is considered good for lactating mothers and very effective against stomach problems. It is typically cut into small pieces, washed and cooked as *Bhaji* with salt, chili, garlic and onion to be served with warm rice. It is also boiled with the rice and separated to make a *Bhorta* (mashed preparation) with onion, chili, salt and mustard oil. This style saves cooking time and is handy for making an early breakfast for the men leaving for the fields. When more time is available, it is fried in a pot with chili, onion, turmeric, salt and garlic and then small fish are added with more ground spices such as coriander. Water is added and the dish is boiled for sometime until a *Jhol* or soupy dish is ready. During the drier winter season *Dondo kolosh* (Leucus aspera) is cooked in a mixture with other greens. The *shaks* derived by plucking leaves from cultivated crops such as mustard, pulses, and potato are also cooked in the winter as mixed *Bhaji*.

Rokeya Khanum Tulu with Bhorta , Charchari, Bhaji and Ghonto. (clockwise from bottom left)

In the mixing of the greens, typically one species is used as the main item and others are added for taste or other properties. For example, *Shenchi shak* will be mixed with smaller quantities of *Kolmi shak* (Ipomoea aquatica) and *Kalai shak* (Derris trifoliate) and mixed together with onion and green chili. Sweet potato, cultivated for its tuber, produces a derived green cooked with *Botua* (Chenopodium album), *Shanti, Dondo kolosh* (Leucus aspera) and several other leafy greens. Our discussions with women found consistent combinations of greens across households, in patterns of two, three, four and five species.

Women know very well the tastes and medicinal qualities of different *shaks* after mixing and individually. The main criteria for combining particular species is based on a local classificatory system which divides leafy greens into sweet, sour and bitter types. Bitter types are usually of medicinal value while the sweet and sour types have particular food values.

Mixing of these types is based on specific knowledge of the properties of each plant, their combined effect and the particular circumstances of the household. Ideally, the bitter *Gima shak* is not mixed with a sweet shak because mixing reduces its medicinal qualities. It is eaten in late March and early April to strengthen the body against diseases of the coming monsoon. But it may be mixed with vegetables such as brinjal or eggs so that children can tolerate the bitter taste and yet benefit from its medicinal qualities. There are other *shaks* such as Jute leaves or *Kanai shak* which become sticky after cooking with water and are therefore not mixed with other types. *Netapeta* is a sweet tasting *shak* cooked with *Henchi* (a bitter *shak*) and *Amaranthus* (a sweet *shak*). Sometimes, one *shak* is taken as the major *shak*, and then combined in lesser proportion with other *shaks* of different tastes.

Shak have definite eating times. In Tangail, *shaks* are usually eaten two times a day, either in the morning and afternoon or in the afternoon and evening. The afternoon meal contains a larger amount and great diversity of *shak* compared to the other two times of day. This is partly because larger amounts of *shak* can be collected during the day while working in fields or visiting common lands, and they are freshest for the afternoon meal. They are also digested more easily when eaten during the day, as opposed to evening time. The morning *shaks* are usually collected from immediately around the homestead, typically in a matter of minutes before the morning meal is prepared. A survey of households in Tangail generated a list of 17 *shaks* consumed mainly in the morning, 42 in the afternoon and 19 in the evening.

In general, the poorer families consume a greater diversity of leafy greens, usually prepared as *Bhorta* (mashed preparation) and *Ghonto* (water preparation). They tended to economize by not using oil in their preparations and few spices other than except chili. By contrast, the better-off families usually cook their greens as *Bhaji* or *Charchari*, preparations which require more spices and oil. They also consume a lesser diversity of species.

Uncultivated Greens – the Nutritional Values

Nallakashe.

It is common knowledge that leafy greens are beneficial to people's health. The broader context of dietary diversity and its relationship to human health is also apparent but widely ignored. The shift from a dietary pattern based on diverse sources of nutrition including fruit, vegetables, animal protein and grains to a diet high in saturated fat, sugar and refined foods and low in fiber is virtually complete in North America and Europe, and is occurring very rapidly in middle income countries such as Mexico, South Africa and Egypt. It is also widespread and increasing in the urban populations of even the poorest countries in the world. What is troubling is that the "nutrition transition" is directly related to rapid increases in the incidence of nutrition-related non-communicable diseases such as diabetes, cardiovascular disease and certain kinds of cancer, diseases that are now the main causes of disability and death globally. Research suggests that this is a looming public health crisis of enormous proportions.[1] Only a few countries such as South Korea have been relatively successful at directing the nutrition transition and avoiding the concomitant increases in nutrition related non-communicable diseases by promoting traditional low-fat, high-vegetable local cuisines that include many so-called "wild greens".[2]

Thummikura.

When the farmer women of Medak district talk about the food and nutritional qualities of uncultivated greens what emerges is a regular rhythm of consensus. The expressions to describe them range from utilitarian ("it helps us cope with the poverty") to unabashed admiration ("Doggali greens are tastier than chicken eggs"). They are also unanimous in the view that uncultivated greens bring "plentifulness" (*barkat*) to their meals.

Pullakoora.

This is particularly true when the greens are cooked with lentils, cowpeas, *mung* beans or other pulses as the total volume of the dish increases, making it possible for the entire family to eat a complete meal with few cultivated ingredients. Other dishes using uncultivated plants are curries (with or without other vegetables) and soups made from pigeon pea or horsegram. These are typically combined with sorghum bread (*chapatti*) or one of several different types of millet grains, along with a hot sauce made with hibiscus pods or a condiment made from ground sesame seeds. Thus, while the cultivated landscape provides carbohydrates and proteins in the form of grains, pulses and oil seeds, the uncultivated landscape completes the diet with the micro-nutrients contained in the greens.

Pressing samples for laboratory analysis.

Some 79 different uncultivated leafy green vegetables are used by the farmer women of Medak as food.[3] Many are extremely rich in micro-nutrients important to human health.[4] For example, *Nallakashe* (Solanum nigrum) is an outstanding source of vitamin A (23417 IU) while *Jonnachemchali* (Coculus hirsutus) is very rich in calcium (3237 mg/ 100g), Zinc (2.7 mg /100g), magnesium (520mg/ 100g) and manganese (21.3mg/100g). *Yelakachevula Koora* (Merremial emarginata) is rich in copper with 0.9 mg/100g and phosphorus with 166mg /100g. Many of the greens are rich in Iron, including *Pullakoora* (Oxalis corniculata) 139mg/100g, *Jonnchemchali* (Coculus hirsutus) 111.3mg /100g, and *Thummikura* (Leucas aspera) 81.6mg/100g. These leafy greens, along with the pulses and grains of the Deccan, are helpful in the prevention of Iron Deficiency Anemia (IDA), a condition especially harmful for pregnant women and their babies.

A recent study found that the nutritional values of many of the leafy greens in farmers' fields are higher in the relatively infertile red soils characteristic of the mixed farming systems of the poor compared to the more fertile black soils farmed by better-off groups.[5] Assessment of vitamin C and beta-carotene content of leafy greens found that half of the species have higher values when grown in red soils compared to black soils. This is a striking observation of genotype by environment interaction, and a direct challenge to assumptions regarding the inherent inferiority of the red soils of the poor.

While laboratory analysis of the leafy greens establishes beyond doubt their nutritional values, other qualities of the leafy greens are identified by local people using a different set of parameters.[6] The feeling of strength in the body provided by consuming greens is associated by villagers with the process of blood formation, and is probably a good proxy for sources of Iron. Taste is also considered. *Pundi*, the hibiscus plant, has been in the farming environment of Andhra Pradesh for centuries and is considered by many to be the tastiest of leafy greens. Food for special diets is a parameter through which women classify the uncultivated greens used as convalescent foods for children, lactating mothers and adults who are recuperating from an illness. The plants ranked highly as diet food also score highly on the parameter of providing strength to the body and taste.

By contrast, foods that score high on medicinal value,

another parameter, are not perceived as capable of giving strength to the body and hence are of little use to convalescing people. They are, however, good for treating common ailments. Some greens are used to prepare an expectorant for children suffering from a cough and cold while others are used to help in the relief of piles. There are greens to treat night blindness, rheumatism and bone pains. Decoction of a particular green is considered an excellent tonic for urinary problems while another is used to relieve menstrual pains and dysentery.

Ease of preparation refers to the amount of labor involved in the collection, separation, cutting and cooking of uncultivated greens. A green such as *Pulchinta* is given a low score on this parameter because the small, tender leaves require painstaking separation leaf by leaf from the fibrous stem. Storability is also considered, and the few greens available with this quality are highly valued. *Attheli*, for example, is dried in the sun and stored for use later in the year. As many of the poor do not have land around their homes and must walk some distance to an agricultural field or village commons to collect uncultivated greens, having stored greens on hand is a treasure.

Market value was also recognized as a relevant parameter but one which applies only to a few of the leafy green vegetables. A handful of species (for example *Kothmir* and *Pudina*) are used in virtually all curries to enhance the taste of the dish. They are in high demand and consequently have good value in the market. Most other greens do not currently have a monetary value because they are abundant and readily available.

Entrance to "Café Ethnic", a DDS enterprise dedicated to serving Deccan foods.

Self-regeneration is a quality for which uncultivated foods have to stand the test as they are not expressly planted by human beings. Half of the species ranked were given a very high score for their ability to regenerate themselves because they germinate easily and are accessible either in dry lands or irrigated areas. Some greens were given low scores on this parameter because they require special care and attention at different stages of growth.

Barkat is a unique local concept which expresses whether or not a food cooked with another adds bulk to the diet. On this parameter many leafy greens rank highly. Narsamma from Kalmela village brought this concept alive with the observation that "a whole family can continue to live on a green like Pundi".

[1] Popkin, B.M., S. Horton, S. Kim, A. Mahal and J. Shuigao. 2001. "Trends in diet, nutritional status, and diet-related noncommunicable diseases in China and India: the economic costs of the nutrition transition." Nutrition Reviews 59:379-390. Also see Johns, T. 2002. Plant genetic diversity and malnutrition: Practical steps in the development of a global strategy linking plant genetic resource conservation and nutrition. African Journal of Food and Nutritional Sciences. 3: 98-100.

[2] Lee, NJ, B.M. Popkin, S. Kim. 2002. "The unique aspects of the nutrition transition in South Korea: The retention of healthful elements in their traditional diet." Public Health Nutrition 5(1A): 197-203.

[3] The diversity in the fields and meals of the rural people in Medak is considerable, including some 317 different food items: 27 food grains, 79 leafy greens, 39 roots, tubers and vegetables, 18 spices, 27 pulses, 18 nuts and oil seeds, 9 drinks, 50 fruits, 39 animal foods and 11 miscellaneous foods. Documentation of this diversity and its nutritional qualities was facilitated by S. Yesudas and P.V. Satheesh, in collaboration with H. Kuhnlein of McGill University (http://www.cine.mcgill.ca/documents/manual.pdf) and M. Schmid (Traditional food consumption and nutritional status of Dalit mothers and young children in rural Andhra Pradesh, South India. 2005. PhD Dissertation, McGill University, Montreal, Canada).

[4] Deficiencies in micronutrients cause a number of important conditions including blindness and diarrhea (due to vitamin A deficiency), anemia, poor motor and mental development, complications during pregnancy and delivery (due to iron deficiency), lethargy, goiter, and extreme mental retardation in newborns (due to iodine deficiency).

[5] Report of the National Institute of Nutrition (Hyderabad), for the Deccan Development Society. 2004. See www.ddsindia.org for details.

[6] The parameters and scores assigned by people were identified by 12 women from different villages ranking 40 different uncultivated leafy green vegetables. Reddy, S and S. Yesudas (compilers). 2000. A Participatory Rural Appraisal on Foods by Eedulapalle Bayavva, Hoti B Sangamma, Basantpur Santhoshamma, Raikode Balamma, Pastapur Pushpalamma, Metlakunta Paramma, Humnapur Kashamma, Pyalaram Thukkamma, Kalmela Kamalamma Narsapur Bagamma and Kalmela Narsamma; Deccan Development Society, Pastapur, India.

Sama Kichidi

INGREDIENTS

Little millet	1 kg
Water to cook	2 liters
Greens	50 g
Onion	25 g
Greengram dal	250 g
Green chilly	6 no.s
Oil	20 g
Seasoning seeds	
Curry leaf	
Ginger garlic paste	To taste
Salt to taste	

Cooking time: 15 minutes

Method:

- Wash and chop onion, chillies and greens.
- Clean and wash green gram dal and soak for half an hour.
- Take a big clean vessel and heat the oil.
- Add the seasoning seeds (mustard and cummin seeds), curry leaf, chopped onion, ginger garlic paste, greens and green gram dal.
- Allow it to fry well, add salt to taste and add water (two liters of water to 1 kg. little millet).
- Cover the vessel with a lid and allow to boil .
- Wash little millet and drop it in the vessel
- Stir well, cover with a lid, and cook to a firm consistency on a steady flame.

Nutritive Values: For 100 g of the Ingredient

Ingredient	Proteins gms	Fat gms	Minerals gms	Fiber gms	Carbo Hydrates g	Energy mg	Calcium mg	Phosphorous mg	Iron mg
Little millet	7.7	4.7	1.5	7.6	67.0	341	17	220	2.5
Green gram dal	24.5	1.2	3.5	0.8	59.9	348	75	405	3.9
Oil	-	100	-	-	-	900	-	-	-

Special Features

- All in one Dish

Note : The same recipe holds good for Foxtail millet Kichidi, Barley Kichidi

Sama Kichidi, a traditional recipe of little millet and greens.

Collected greens such as Dheki sak (Dryopteris filixmas) add taste and nutrition to the daily diet.

Understanding the dynamic relationship of biodiversity and local food systems allows us to question the inevitability and desirability of a transition to diets based entirely on cultivated diversity. Increases in the production of so-called "stable crops" can only be considered an increase in food production if they really make a net contribution to the availability of food at the household level. This is an empirical question and needs to be grounded in the context of specific cultures and ecosystems.

The role of uncultivated foods in addressing the regular food and nutritional needs of rural communities was assessed for Tangail, Bangladesh through a dietary intake survey in 10 villages.[1] Some 50 families from a stratified random sample of 2,285 families were interviewed, representing four class categories defined in the National Census by farm size: very poor (20% of the total population), poor (44%), middle (32%), and relatively well-off farmers (4% of the total population). The very poor families did not own any cultivable land while the largest farm was 4 acres. The vast majority of the farmers in the sample villages are Nayakrishi farmers practicing mixed cropping and crop rotation without the use of pesticides. This provided a context where uncultivated plants and fish were likely to be relatively abundant on the farm and safe to eat, compared to villages where monocropping and pesticide application is common.

Data collection was undertaken in late October and early November, a time of the year when food, including rice and cultivated leafy vegetables, is relatively abundant. The study period was a normal winter season without significant stress conditions due to floods, drought or other calamities. Consequently, any observed use of uncultivated foods such as leafy vegetables during this period was probably due to dietary choice, not because families had no other option. Nevertheless, limited access to other foods is a factor in individual cases, especially among the very poor.

To collect the data two UBINIG researchers stayed in the study villages and took direct observations of dietary intake from early morning until the evening meal had been completed. In each household food consumption

was monitored for three days and the ingredients used in each meal weighed with a scale before and after cooking. Uncooked items were weighed after they were washed, cleaned and drained of all water, as is normally done in cooking. The cooked food was weighed as it is usually served; in the case of curry it was weighed with the gravy (*jhol*).

The uncooked ingredients were divided into two categories: cultivated ingredients (mainly rice, wheat, pulses, meat, large purchased fish, eggs and cultivated vegetables) and uncultivated ingredients (mainly leafy greens and tubers, as well as small collected fish). This division was possible because the items are easily identified and often cooked separately. The observations for 3 days of meals were totalled and the proportion of uncultivated foods in the diet was calculated based on the weight of uncooked ingredients.

The Very Poor

Sufia Begum, a woman of 33 years in the village of Maushakathalia, has no cultivable land and no economically active men in the household but has a small homestead she shares with a daughter (14 years) and a son (8 years). This profile is typical of families in this economic category. Only one very poor household among those surveyed had a cow, although many owned chickens or goats. Food consumption estimates for the study period indicate that families in this category rely on uncultivated foods to supply a high percentage of their food by weight, ranging from a low of 38% to a high of 89%, and averaging some 65% across the families surveyed. They also met virtually all of their fuel and fodder needs from uncultivated sources. This category of family also sells or exchanges uncultivated foods for other food and goods they need.

The Poor

Hazera Begum (52 years of age) is somewhat better off than Sufia because she has both homestead land and a small amount of cultivable land (less than 1 acre). She also owns a cow. She is a widow and has two small grandchildren living with her. This level and type of assets is typical of the poor families surveyed. On average, they derive about 55% of their food by weight from uncultivated sources. Discussions with them indicate that 70% to 80% of their fuel and fodder needs are also met from uncultivated sources.

Middle Families

Saleha Begum has about 2 acres of cultivable land and a

small homestead, the land assets of middle families. She owns two cows. The household is comprised of herself, her husband, a grown son and a daughter-in-law. On average, families in this group rely on uncultivated food sources for about 52% of their food by weight. Discussion with these families indicated that uncultivated sources provide about 30% of their fuel and fodder needs. While the land assets of families in this category would seem to provide them

Mola, Darkina and other small fish.

with economic options, uncultivated sources are still very important to their daily diet.

Better-off Families

Abdul Jalil, his wife, two sons and a daughter have a little more than 3 acres of cultivable land and an ample homestead as well as several cows and poultry. Despite these assets the family still relies on uncultivated food sources for 43% of their food, with an average for this family category of approximately 34%. This is a remarkably high percentage for a food source virtually absent from the formal marketplace. This type of family also draws on uncultivated sources for about 20% of their fuel and fodder needs.

The survey indicates that for all social classes in the villages uncultivated food sources make up a large part of the daily diet during a time of the year when cultivated food sources are also relatively abundant. Far from being a minor supplement or simply a crisis food, the leafy greens, tubers and small fish collected by people from the lands and water bodies of their communities are a vital part of their daily diet. While this was not surprising in the case of the very poor, what we had not anticipated prior to the study was that even middle and better-off families draw significantly on uncultivated sources for their food, feed and fuel needs.

Where biodiversity-based ecological farming practices are dominant these food sources are available and systematically incorporated into the daily diet. This suggests that a shift to ecological farming in other regions would enhance access to uncultivated foods, ensuring a large and strategic portion of the food needs of community members, regardless of social class.

[1] *The villages included in the study are all in the Atia Union: Hinganagor, Goaria, Kandapara, Nanduria, Binduria, Khathalia, Shailkhai, Garasin, Mamudpur, Baroi Atia.*

Cultivating the Social Landscape

Survival Strategies of the Very Poor

Mother and daughter together collecting sugarcane roots for fuel.

The importance of uncultivated food sources in the survival strategies of the very poor is quite well known in the villages of Bangladesh. But what is the nature of this contribution? No one can have a really clear answer to this question. Collecting information on how poor people survive is not easy. It is even difficult to find them at home when we go looking for them. In the village when we ask, "Where are they?" the answer that comes back is "Chak", meaning they are in the open fields or "out there", collecting her family's daily nourishment from the road side, government lands or from the private lands of others.

Near the end of the Aman season the very poor are very busy in the sugarcane fields harvesting for farmers. Later in the year they are harvesting potatoes and preparing seedlings for the paddy fields of farmers. They may receive some money for this labour, which they will use to buy oil and salt or pay for school expenses and debt repayments. But they will also collect the straw, which is no longer needed to cover the ground in the potato field, and bring it home for fuel. They will take potatoes in partial payment for their harvesting work. They will pick the jute leaves in the farmers' field for food and collect the uncultivated leafy greens along the side of the rice field. They will sell eggs from their ducks or chickens to buy rice. This is their livelihood.

Standard methods of poverty measurement cannot explain how these people survive. To fully appreciate the survival strategies of the very poor we need to go beyond the calculation of food expenditures at the household level, the main focus of conventional studies. We started by interviewing 10 women from 5 villages.[1] All were landless and either widows, divorced or living with husbands too sick to work. Families ranged in size from 2 to 6 members and ages between 30 and 70. All would be considered "hardcore poor" under standard definitions of poverty.

We did not use a questionnaire except to ask what they did on a daily basis for each month of the Bengali calendar, starting from *Baishak* (mid April - mid May) up to *Chaitra* (mid March to mid April). We arranged to sit down with the woman in a quiet place so the discussion and flow of information would not be interrupted. The use of the Bengali calendar and the emphasis on food sources and other means of daily living made it possible for them to tell us about the availability of each kind of uncultivated food and indicate their own use of those foods. It took a long time to interview the women and we had to check the information again and again. In this process of rechecking the women also helped us a lot. The discussions proved to be very interesting and revealing of many unknown aspects of their livelihoods.

fields that were also reported to be collected for food and fodder.

Fish are available throughout the year, except during the 4 months of the dry, winter season. The Mola, Icha and Chata are three kinds of fish that are abundant during the rainy season, even in small water bodies such as under bridges and in culverts. During the monsoons, fish such as Shing and Koi are also available in the paddy fields.

The fruits of some trees such as *Jambura* can be picked by any child from the village, even when they are on the homesteads of a different family. While not a part of regular meals, these snacks help to meet the nutritional needs of the children.

The uncultivated food sources mentioned by the women form five major categories: leafy greens, fish, fruit, roots and tubers, and secondary parts of cultivated plants. These items are mainly used to meet the immediate food needs of the families, although some, such as snails and rice straw, are used as feed for livestock or fuel.

Being able to feed goats along the roadside or field edge helps ensure a source of cash.

Leafy greens are particularly useful because they are available throughout the year and, when combined with fish and rice, make a balanced meal. Some species such as Helencha and Netapeta are available all year round while others are only eaten seasonally. There are many partner plants such as *Batua* and *Dondokolosh* in the cultivated

Roots and tubers are collected most months of the year, mainly from the water bodies. Children do much of the collecting, and report that they consider it fun to do. *Shapla* is the flower of the Water Lily. It has an edible nodule at the edge of the root. *Dhep* is the grain-like substance that grows in the flower of the Water Lily and is eaten in puffed form like rice. Children like the green Asnaru mixed in with mollases, and Hindus use it for worship. During the two dry months of *Baishak* and *Jaistha* no roots or tubers are available but they are abundant during the rest of the year.

Children also collect the snails as feed for the ducks. Snails are known to be very good for duck egg production, and are available about half of the year and very abundant during the month of *Bhadra* (August –September). They also collect a special kind of grass called *Khudi pana* which floats on the water. It is a very good feed for ducks.

The secondary parts of cultivated plants are collected mainly during the winter months and dry period of *Chaitra* (March-April) and *Baishak* (April-May), and include the fresh leaves of potato, mango, wheat, barley, turmeric, banana, jackfruit, gram, pumpkin, jute, sweet potato, sweet gourd, onion, garlic and beans. These are collected without damaging the growing plant or affecting production of the main crop. Common areas and farmer's fields are the most significant sources of the diversity used by the women, underlining the importance of social relations which maintain these spaces and enable access by the poor.

Poor families do not only consume what they collect; they also exchange these plants and animals with others to get whatever they need. Exchanges are most common between the women of the poor and the better-off families. Women in the better-off farming families do not go out to collect the leafy greens they want to eat but rather get them from the

If a mother has a daughter, both will go to collect greens. This flower is used to make bora, a dish of greens with lentils.

poorer women, sometimes on request for a particular leafy green. These transactions typically don't involve cash, but rather other products such as rice, broken rice or rice husks used as feed for chickens. Sometimes they will exchange for paddy straw used as fodder for the cow. This enables the poor women to keep hens and ducks, and even a cow as an income-earning activity. Interestingly, in this exchange there is seldom any bargaining or estimation of how much is given for what. The poor woman brings the desired leafy green, and the better-off woman gives her something in exchange needed by the poor woman.

Both cultivated and uncultivated items are sold to get money to purchase other household goods such as soap, oil, salt, spices and clothes. Poor families grow a few vegetables on their homestead lands and collect uncultivated plants for sale in the local markets. There is demand in the local markets for certain leafy greens like *Kolmi, Dheki, Kochur loti* and aquatic tubers like *Shapla, Shaluk,* and *Dhep*. As these exchanges are informal and depend a lot on what others want to give, the poor only sell in the market if they are desperately in need of money.

Women sometimes refer to the Jackfruit (Artocarpus heterophyllus) as the "rice tree", in recognition of its vital role in times of food shortage.

The poor regularly collect fuel for cooking. Very little of what they collect is actually "wood" from trees, but rather materials that are cast off from other plants: leaves, straw, roots, dead branches of trees and parts of plants from around the village. They spend a lot of their time during the day on this activity, as without it they cannot cook. They also actively store fuel for use during the monsoon months of *Ashar* (mid June to mid July) and *Sraban* (mid July to mid August), a time when there is often a fuel crisis among many village families. The poor women who collect and store fuel in the months leading up to this period are able to sell it during the crisis months. This is a direct contribution to their communities.

Collection for fuel is done in a way that helps clean the field for the next crop: roots of paddy, sugar cane, wheat, and other crops are dug up from the fields and transformed into fuel. Dry bamboo leaves, leaves of mango and jackfruit trees, paddy straw, dry banana tree stems, jute sticks, betel nut shells, roots of bamboo, dry okra plants, leaves of sugarcane, dry sugar cane, water hyacinth, wheat straw, and husks of mustard are also collected for fuel. Some of the fuel sources, such as *Moilta* and the *Bait* tree, grow on the roadsides. They regenerate very quickly when cut, and are carefully managed by the women who depend upon them. Cow dung from the cow shed is usually saved for manure rather than for fuel, but cow dung found on the roadsides and near the homesteads is turned into fuel. Stalks of jute and other plants are covered with the cow dung and dried to make a hot burning fuel, for home use or sale.

The month of *Bhadra*, in the middle of the monsoon, is seen as the best month of the year for the poor as there is a lot of food, feed and fodder available from uncultivated sources. This is because during the monsoon diverse leafy greens are abundant in common areas as are plants and fish that grow in water bodies. Exchanges of small fish and leafy greens for other foods increases as well during this period. This situation is the opposite of farmers that depend mainly on the harvest of cultivated crops for survival. Cultivated vegetables are particularly scarce during the monsoon as much of the land is under water or rice cultivation. This is a period when the risk of micronutrient deficiencies is greatest for vulnerable populations such as women who are pregnant or menstruating and young children. The uncultivated food sources of the very poor, rich in vitamin A, iodine and iron, are seasonally strategic in helping to meet these nutritional needs.

The picture from the monthly accounts the women provided suggests that they depend on the uncultivated and cultivated biodiversity available to them in the local environment and through relationships with others in the community for almost 100 percent of their livelihood. The policy implication of this observation is enormous: food and livelihood security for the very poor, a population outside the reach of conventional development schemes, is enhanced by taking care of the social landscape where uncultivated plants and animals can be collected, exchanged and safely consumed.

[1] *The villages of Nanduria, Moushakthalia, Kathalia, Gopalpur, Aloatarini, Goaria, Delduar and Kanadapara in Tangail District.*

Manure applied to a jute stalk makes a valuable fuel.

Poverty and the Social Landscape

Our research with communities in the Deccan and Bangladesh on the role of uncultivated foods in the food systems of the poor provides several important insights into the nature of their survival strategies and what sustains them. It suggests that there is a critical relation between survival strategies, ecological and biodiversity-rich production systems and the customary rights of the poor to collect and gather food from their surroundings. These ecological and ethical relations fashion a "social landscape" that ensures survival, no matter how difficult that may be.

These findings have important implications for the design of poverty alleviation programs, most of which have been abject failures for the poorest of the poor. The so-called "hard core poor" are typically out of reach even with micro-credit schemes designed with these populations in mind.[1] For some economists this phenomenon is evidence of a "mismatch" between national or macro-level growth strategies on the one hand and micro-level poverty alleviation schemes on the other.[2] Even with economic growth the programs do not have the capacity to address the vital survival issues of the very poor.

In our view, the failure of poverty alleviation schemes is due to an overemphasis on income and employment initiatives and a profound disregard for expenditure-saving activities (self-provisioning) and livelihood strategies not related directly to employment. Rural economies in places like the dry lands of the Deccan and the flood plains of Bangladesh depend not only on the production of field crops for sale and the employment of farm laborers but also a wide range of productive activities not mediated by the market. Rural people collect medicinal plants in common and private spaces to address health problems, use crop residues to feed their animals, exchange services with trades people and crafts people in the community, collect forage from uncultivated lands and forests, collect fish in open and closed access water bodies, collect food while weeding crops for neighbors, and so on. These are the real survival strategies built up by rural communities over the course of human history, bringing them into constant interaction with many inter-related natural and social resources.

Unless governments can absolutely guarantee significant new rural or urban employment for the very poor, something that has been impossible to achieve for decades now, it would be irrational and unethical to continue to sideline or destroy these strategies.

We acknowledge that further research is needed to substantiate and enlarge on our observations regarding the ecological and ethical basis of the productive activities

Women's knowledge must be acknowledged. Shahera Begum of Bangladesh.

and survival strategies of the poor. On the one hand, it is important to avoid a romantic notion that traditional community institutions are always equitable and that traditional agriculture is able to adjust rapidly to changing circumstances. The very poor suffer enormously, and often face severe exploitation at the hands of local people. On the other hand, a deterministic notion that individuals and communities are helplessly caught in a poverty trap beyond their control, and can only be rescued from their plight by the technology of corporate agriculture, is equally unacceptable.

Policy challenges raised by recognizing the complex interaction of cultivated and uncultivated biodiversity include the urgent need to remove threats to ecological and biodiversity rich production systems posed by pesticide use and intensive monocultures, and to protect public spaces and social institutions that enable the poor to access food and livelihoods where they live and work.

To this defensive strategy must be added a range of measures to intensify the further improvement of ecological and biodiverse farming practices and to strengthen the capacity of people as individuals and as communities to

Helping a farmer deepen a well in Basantpur, India.

assume the responsibilities and rights of governance over the rules and institutions that regulate economic activities. This would involve collaborative decision making between governments and their citizens concerning interventions that transform public spaces such as roadside development, floodwater management, and reforestation schemes.

Fragmentation and enclosure of the social landscape is a broad policy trend that has enormous direct impact on the livelihood options of people who depend on these spaces to graze animals or collect items for food or sale. This is particularly serious for women and children. Village pastures, roadsides, pathways, so-called waste lands, village ponds, community forests, rivers and backwaters are critical spaces for women's livelihoods and children's nutrition. These feminine spaces are being fragmented, privatized and redesigned in ways that reduce or eliminate the customary rights of the poor to gather resources and fashion their livelihoods. Suraiya Begum, an elected member of the Delduar Union in Bangladesh, protested in a public meeting against the planned construction of a road in her village on the grounds that it would destroy the sources of fodder for the goats raised by the poor women. She also objected to the selection of trees to be planted along the roadside because they would not bear edible fruit for the benefit of children in the village. She was shouted down by men in the meeting who censored her for opposing "development" in their community.

The significance of a social landscape is immediately evident to village women such as Saleha Begum, a 35 year-old mother of two responsible for feeding herself and her children while her husband is away working as a laborer. The average income of a laborer in Tangail is only Tk.1000 to Tk.1200, an insufficient amount to purchase daily food and other immediate needs of the family. Consequently, she collects a variety of leafy greens from cultivated plants in neighbouring fields including jute leaf, pumpkin leaf, white gourd leaf, the tender leaves of potato, *kolmi, helencha* and the tender leaves of *bon kochu*. On the way back from school her children collect some small fish from a backwater and also bring snails for the ducks they are raising. She cooks and serves the fish and greens, an immediate and healthy meal from nature. "You do not need to spend money for the uncultivated food. It is available", says Saleha. Throughout Bangladesh, these "private" cultivated spaces (farmers fields) and "public" uncultivated spaces (a backwater) provide food near at hand any time it is needed.

Field edges, fallow strips and common areas constitute a feminine space.

Ask any landless Dalit in Medak where she went to collect her day's greens and the answer comes back, "Baayila kaada" - from around the irrigated lands. None of the very poor are likely to own an irrigated patch of land. From the answer it is clear that she has gone to some farmer's irrigated patch of land to gather her day's greens.

Communities of this nature draw on the biodiversity of local spaces: fields, field boundaries, seed stores, household patios and common areas. They also rely on the social and institutional relationships that regulate access to biodiversity: gender relations, community membership, cultural practices, and specific legal provisions. Space in these cases is not just marked physically, but rendered complex by ethical relations and ecological dynamics.

Dalit women in Medak are highly dependent on wage work, typically dedicating about five months a year to this activity: three months during the rainy season and two months in the winter season. Every time a group goes to a farm for weeding, they have the right to carry back all the plants they have weeded out. The plants may be greens which are cooked for human consumption or grass for cattle consumption. Either way, the workers have the right to plants they have taken from the soil. Unfailingly, when they return home, they come back with uncultivated greens and fodder as byproducts of their weeding operations. If they are hired for sugarcane harvest work or for sugarcane crushing work, they again have an automatic right to collect and take back home a head-load of cane tops, which is a delicious green fodder enjoyed by the cattle at home.

During the monsoon months, when a lot of grass grows in the empty patches of the farm, landless cattle owners can get special permission to take their cattle inside a farm and let it graze under supervision. Such supervised, careful grazing of animals is also permitted all the time for small ruminants like goats. It is a common sight in the Deccan to see women go with one or two goats when they leave home for weeding in someone else's field and graze them in a limited area while they are working. On single cropping farms, after the harvest is over, the land turns into a common property for herders. Until the new crop is planted, anyone can drive their cattle into these farms and let them graze the crop residues. The farmer is also happy with this arrangement because the cattle deposit their urine and dung on her or his farm, thereby increasing the fertility of the soil. Some studies also point to a deeper ecological relationship in this arrangement. If the crop residues are left unattended on the farm, in the hot climates of the Deccan they can easily be targeted by termites which in turn can pose a number of hazards. In the scenario where cattle graze, digest and excrete the residues, the conversion into soil-building nutrients is catalyzed. This is a symbiotic relationship where both the herder and the farmer derive significant mutual benefit.

Access to grains is also ensured through social contracts between farmers and artisans of the village. The village potter, carpenter, blacksmith, washers, weaver and other service groups enter into precise arrangements with the farmer for their grain needs. For the pots provided for ritual and other purposes, the potter gets a part of the harvested grain. So does the carpenter and blacksmith for the farm tools that they have created or repaired for the farmer. The weaver who supplies ritual clothes at the time of birth, marriage or death in the family is entitled to a share of the harvest. And the washer has her share for all the family clothes she washes throughout the year.

In each of these cases, their needs are better served by farmers who are not producers of single grains but rather of mixed crops. These farms are a ready-made supermarket with a range of food items one needs for good eating (nutritious eating, tasty eating, varied eating, culturally satisfying eating). Farmers in turn have access to the agricultural implements and services they need and ready local markets for their surplus produce.

Apart from all the most visible service castes there are many subaltern groups in the complex social web of the Deccan who benefit from barter systems that serve as a kind of social safety net. For example, the Talaris supply tobacco leaves every day for farming castes while the Tammalis

supply flowers for special occassions. These castes are dependent on two different diversities which sustain their livelihoods and relationships in the agrarian community.

The castes on the margins are also a part of the barter system. The Begaris are the grave diggers for the community and the Madigas do the "undoable" jobs like disposing of cattle carcasses. People belonging to these communities have gleaning rights for their invaluable services. They come to the field the day after the harvest is over and go straight to the threshing floor. Any grain leftover on the threshing floor that has not been bagged is theirs, no questions asked. They are also entitled to go round the harvested fields collecting fallen grains on the field, an activity that can easily result in 30-40 kilograms of grain over a few days. The Bichapollu, the mendicants (beggars) of the village, also are entitled to a share of the harvest. Over two to three weeks during the various harvest times groups of people can be seen moving from field to field collecting their entitlements.

These forms of access to food and livelihood point to a kind of "private commons" available to the rural poor. The extended work relations between bigger farmers and farm workers and among different occupational castes keep certain obligations alive. One such obligation is that

Removing stones from a field makes a contribution recognized across caste divides.

" At the time of drought, we ate Nakka Thoka (seeds from a wild grass resembling fox tail millet) with the leaves of Doggali and chilies. I gathered about four bags of seeds from a field. "

Sangamma, a woman farmer from Hoti-B village, Medak District

when a poor person enters your field to gather uncultivated foods, you do not turn her or him back. The boundaries start vanishing in a sharing atmosphere. Any village person can walk into any private field to gather uncultivated foods for their daily consumption, whether they are greens or fruits. In return, when the land-owners are in need, the villager will help. For example, if there is a wedding in the house, the poor will come and clean the house, paint it, cut fuelwood, help in cooking -- be of assistance in a myriad of ways. Access to the fields is in return for such favors done in the past or anticipated in the future.

In times of hardship, when the crops die on the land due to drought or excessive rain, the private property distinction gets completely demolished. And the use of uncultivated foods increases proportionately. The Medak region had the traumatic experience of going through a severe famine 20 years ago. The memory still haunts people and when they narrate their survival strategies, uncultivated foods occupy center-stage. People survived for four months by eating only the uncultivated greens, in curries along with small amounts of rice or mixed into scarce sorghum flour to make rotis. While doing paid jobs such as well-digging and restoration, they collected the greens from nearby sugarcane fields.

There are many more sophisticated and complex layers to these survival systems, better seen as social safety nets and community builders. At their core is a food production system that is biodiverse, ecological and local, in stark contrast to monocultured farms that have a fragile and vulnerable relationship to the market and are delinked from their communities and the local ecosystem. This base makes possible much untapped potential for incremental improvements in livelihoods, building on what is already in hand.

For example, the Nayakrishi Andolon in Bangladesh have recently experimented with an alternative to micro-credit that builds on the customary system of barter known as "adhi". Very poor female-headed households are given an animal to raise until it has a calf or kid. The woman can use the milk, either for her own consumption or to sell on the market, and keep the animal's offspring as her own to raise. The original animal is returned, and then passed on to another poor family in her village. Maintaining a few animals in the ecological village farming systems of the Nayakrishi is not difficult for women in this situation as plentiful fodder is available from uncultivated sources. Landowners welcome the increases in livestock numbers because this form of controlled, itinerant grazing leaves the dung directly on the land where it can be put to use in the cropping system.

Contrast this modest but tangible development intervention with the schemes contained in the 2020 Vision of the Government of the State of Andhra Pradesh, India's fourth most populous state with some 70 million people, including the Dalit of Medak. The modernization policy calls for displacement of 40% of the rural population from agriculture, to make way for large, mechanized and chemical-intensive farms. It also calls for the reorientation

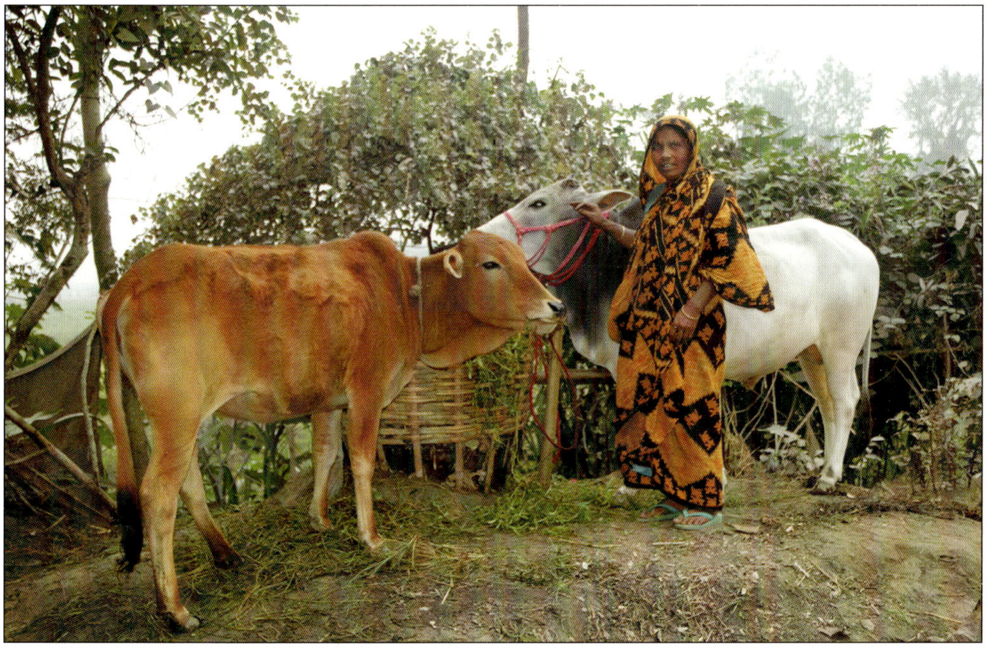

The loan of a pregnant cow is a customary credit system.

The Katkari, a so-called "Primitive Tribe" of the Western Ghats, have been displaced from their lands and are now among India's most vulnerable.

of production for export markets. In this future scenario it is not sufficient for agriculture to simply feed the population it must also do its share to generate foreign exchange. Agriculture, with all its cultural, ecological and survival dimensions, is reduced to a trade issue in the context of macroeconomic planning. The fate of the uprooted families is left to chance on the open labour market.

Disruptions caused by policies of this nature leave rural people with no sustaining link to their homes and their communities. The separation from food sources and the social landscape is inherently conflictual and destabilizing because it leaves people vulnerable to injustice and violence. Whether it drives people from rural to urban areas or demoralizes food-producing communities, the connection to regional tensions is direct and immediate. Migration within and between nations of the region is a significant source of political conflict, as is trafficking of women and children displaced from their homes. Never before in history has the significance of biodiversity in local food systems been so paramount to redefining strategies for peace and poverty alleviation in the region.

[1] Many studies on micro-credit programs have noted that the very poor are excluded from access to credit by factors such as bureaucratic bias and an inability to repay. For examples see R. Rahman. 2001. Women and Microcredit In Rural Bangladesh: An Anthropological Study of Grameen Bank Lending. Westview Press, Boulder, USA; Rahman, A. and A. Razzaque. 2000. "On Reaching the Hardcore Poor: Some Evidence on Social Exclusion in NGO Programmes". Bangladesh Development Studies. Vol. XXVI, No. 1; Zaman, H. 1996. "Micro-credit Programmes: Who Participates and to What Extent?" Working Paper No. 12, BRAC-ICDDRB Joint Research Project, Dhaka.

[2] For a discussion of the "mismatch" problem see B. Sen, 1998. "Politics of Poverty Alleviation" In R. Sobhan (ed). Growth or Stagnation: A Review of Bangladesh's Development. UPL, Dhaka, Bangladesh.

Ecological agriculture creates a regenerative cycle of livelihoods at the village level.

Despite consensus on the contributions of the Green Revolution to the political and scientific advancement of South Asia, the future direction of agriculture in the region is highly contested. There is clear evidence that self-sufficiency in wheat and rice has come with a host of environmental, social and economic problems.[1] At the same time, there has been progress in the further development and improvement of traditional agricultural systems.[2] Farmer experience in South Asia shows that it is possible to improve livelihoods and increase food production sustainably by making more effective use of composting technologies, enhancing the integration of crops, livestock and aquaculture, refining the principles of mixed cropping and landscape design, enhancing local seed, etc.

University-based research has been trying to catch up with these local initiatives, bringing to bear scientific advances in understanding of ecology and the ecological dynamics of agricultural systems to find ways out of the current problems. The convergence of farmer experience and a new scientific basis for understanding ecological farming suggests that there is considerable scope for rethinking and redirecting the course of technological development in South Asian agriculture.

A new kind of economic research is needed to support policy action that can promote these changes broadly. The economics of agriculture is a strategic element in policy debates and of concern to farmers. Put simply, farmers and economists both ask: what kind of farming is economically viable? In the broader policy context, political leaders are legitimately concerned about the capacity of agriculture to feed non-farming populations in the growing cities and in the countryside, as well as the impacts of particular farming practices on the environment.

While agricultural economics has many tools to help answer these questions, the discipline is challenged by several conceptual and methodological problems. It is generally accepted that a failure to internalize the costs of negative environmental and social impacts (so-called externalities) is a key gap in the conventional approach to economic assessment of agricultural options. While it may not be useful to measure everything, and some aspects are not easily quantified, limiting economic assessment to financial costs and benefits at the farm level is clearly inadequate to a meaningful assessment of economic performance for the community or nation as a whole. Failures of this nature can lead to mistaken appraisals of public investments and prices for commodities that don't reflect true costs.

BATC advertisement of a "reforestation" scheme.

Take for example the production of tobacco, a crop established in Bangladesh during the British period and currently promoted by the British American Tobacco Company (BATC). Aggressive campaigning by the BATC, favourable taxes and public subsidies on agricultural infrastructure rapidly displaced food crops from regionally important food producing areas such as Kushtia in western Bangladesh. Tobacco production quickly drew down the natural wealth of the fertile soils in the area and consumed much of the available fuel wood for the curing process. Currently, tobacco requires six times the amount of fertilizer per acre as any other crop widely grown in Bangladesh. The curing process uses some 40 kilograms of fuel wood for every kilogram of finished tobacco product.[3]

Numerous applications of pesticide are also needed to control various plant and soil diseases. This increases the human health hazards of farming and contaminates both the soil and water. Persistent weeds have developed in many tobacco fields, rendering them unproductive not only for agricultural crops but even for tobacco itself.

Throughout this process farmers became indebted to the company and lost ready access to the markets and production inputs (including traditional seeds) of their earlier food-based systems. As the fuel wood disappeared, leaves of trees and rice straw that could otherwise be used as fodder for milking cows and cooking fuel were also diverted into curing of tobacco. Women and children are normally responsible for tending the fires during the curing process, a task that runs for continuous periods of 60 to 70 hours without sleep. Children of tobacco farmers are typically absent from school during this time, and women's burden is increased over and above their normal chores.

Public policies and projects that enabled the BATC to get a foothold in prime agricultural areas, and that continue to allow for expansion of tobacco into new areas, can be seen as symptomatic of a factory model of production. A factory typically focuses on the production of a single commodity and relies on an uninterrupted flow of external inputs. Furthermore, it is not capable of recycling its own wastes internally.

"Mula" a noxious weed in tobacco fields.

These limitations do not need to be reproduced in an agricultural system, which can create multiple outputs, generate its own inputs (biomass, seed, new animals) and recycle its own wastes (composting, surface water management, etc.). The waste of one process not used immediately as an input for another becomes residual humus – to be broken down by the insects, bacteria and biological processes and later taken up by the plants that produce energy through photosysthesis.

The efficiencies created by synergistic and internally

Diversity is the cornerstone of Susilamma's balance sheet.

reinforcing production systems do not get the attention they deserve in economic assessments. Take for example the 1 acre farm of Susilamma, a 27 year old farmer in Raipally, Medak District. She grows on her farm three varieties of Sorghum, two varieties of Foxtail millet, Finger millet, Little millet, Pearl millet, Kodo millet, three varieties of Pigeonpea, two varieties of Dollichos lablab, Horsegram, cowpea, sesame, niger and two varieties of hibiscus. The produce she gets from this land is a mixture of cereals, legumes and oilseeds, along with a vast array of uncultivated greens harvested during the monsoon from the farm beds. These provide her all the elements for a nutritionally balanced food basket, as well as plants of medicinal value. Beside food and medicine the grasses that grow abundantly on the field boundaries and the leaves of trees planted along the field edge provide fodder for her cattle. All the seeds that she plants have come from her own harvest of the previous year. She has selected the best seeds and stored them effectively by mixing the seeds with cow dung ash, layering them with *neem* leaves and packing them in a tight basket of local palm leaves. Some seeds stay uninfested and viable for planting for over 3 years.

Susilamma uses no chemicals on her farm. All of the soil amendments come from the farmyard manure, household wastes and tree leaves which she composts in her backyard. This traditional method has been supplemented in recent times with her own biofertiliser in the form of vermicompost. The soil fertility of her land improves every year through other internal inputs such as the pigeonpea grown alongside the sorghum. This legume not only fixes atmospheric nitrogen on its roots but also provides ample leaf fall as a green manure. On the border and internal boundaries of her tiny farm there are dozens of trees such as glyricidia, pongamia, sesbania grandiflora, perennial pigeonpea and others that enrich the soil and generate useful biomass.

While the health of the soil is her greatest insurance against pests, the enormous diversity on the farm provides a second level of protection. The trees provide a habitat for birds that feed on insects and the absence of pesticides in the system insures that a large number of beneficial insects are also there. Together these measures keep insects well within manageable limits, rarely crossing the threshold to become pests.

Traditional seed storage practices are effective and economical.

The diverse and integrated farms of women like Susilamma are not only highly efficient food producers but also direct contributors to a regenerative cycle of livelihoods at the village level. The fibre produced from the sunhemp and hibiscus in the dry lands of Deccan is an input for local rope makers. The agave hedges are used to produce a very specialized, tough rope for the shipping industry while the more skilled of the rope makers use the fiber to make a host of crafts and carpets. The wood grown as a part of the tree cropping system is the lifeline for rural carpenters. The species are especially apt for making ploughs, carts, and several other agricultural implements in addition to use for doors, door frames, windows and cots for home use. Similarly, potters use the chemical-free soft soil from the farmer's fields to produce specialised pots used for storing seeds and other items. Uncultivated plants sustain a host of human and animal healers throughout rural South Asia and are a key component in the healthcare systems of the Dalit. In this fashion, biodiversity-based and ecological farming

Production of high value mustrd oil.

Entry into the one-dimensional world of economics is not without risk. South Asian farmers have an historically rich experience of agriculture as ecological, economic and cultural processes embedded in the idea of food as a gift of nature, not a product of industry. Food embodies social, cultural and metabolic relations with nature and indeed a kind of language and communication through which social relations are maintained and reproduced within a community and between human communities and nature. Conceptually, the discipline of economics ignores the relations of material 'exchange' in nature and the 'metabolic' social interactions of money and numbers. Food production systems are not assessed as dynamic systems of ecological, economic and cultural organization. The challenge is to bring to economics an understanding of agriculture that is rooted in agrarian histories, cultures and lifestyles and to develop ways of assessing performance that can retain its living character.

systems produce a variety of materials for the village artisan community, expanding and deepening the link between the farming and non-farming social and occupational landscape. While modest, these livelihoods are viable options in their own right and should not be ignored in favour of an idealized and universalized model of industrial employment.

Better research methods are needed to assess the chain effects of technological and production choices both within the agricultural sector (internal dynamics) and across sectors (linkages between agriculture and industry).[3] Similarly, linkages between a range of policies, development goals, and institutions that have a bearing on food systems in a particular context also have to be considered. This requires an actor-oriented perspective on policy-making that can bring to light the key institutional stakeholders, the dominant policy narratives, and the values and interests considered in the decision-making process.

Deciding whose priorities have standing and should be quantified and valued is of particular importance to an actor-oriented assessment of economic performance. Women's perspectives are often neglected, skewing assessments in favor of cash-cropping and other male-dominated domains. Research on economic policy-making in practice needs to develop a language that can be understood and used by farmers, ensuring an effective dialogue between farming communities and other actors. This is needed not only to build agreement on the performance indicators farmers and policy makers want to see measured and taken into account but also to enhance the capacity of farmers to assess their own agriculture and its inherent strengths.

Collecting Date Palm sap for making syrup.

Bamboo basket making is an integral part of farming livelihoods

[1] *The Rice-Wheat Consortium for the Indo-Gangetic Plains comprised of scientists from the national agricultural research systems of the region and the CGIAR is a mainstream source of information on these problems (http://www.rwc.cgiar.org/).*

[2] *Pretty, Jules, and Rachel Hine have undertaken a thorough review of many traditional and emerging agricultural systems around the world, demonstrating the high levels of total farm productivity these systems can achieve. "Reducing food poverty with sustainable agriculture: a summary of new evidence." (2001). University of Essex, Centre for Environment and Society, Occasional Paper 2001-2. Available at http://www2.essex.ac.uk/ces/.*

[3] *Collaborative methods for assessing the dyanmic interaction of elements within agriculture and across sectors in agricultural areas are being developed by the Social Analysis Systems[2] initiative (http://www.sas-pm.com), based on the principles of input-output analysis.*

The Ecological Footprint is 'a resource management tool that measures how much land and water area a human population requires to produce the resources it consumes and to absorb its wastes, taking into account prevailing technology' (www.footprintnetwork.org)

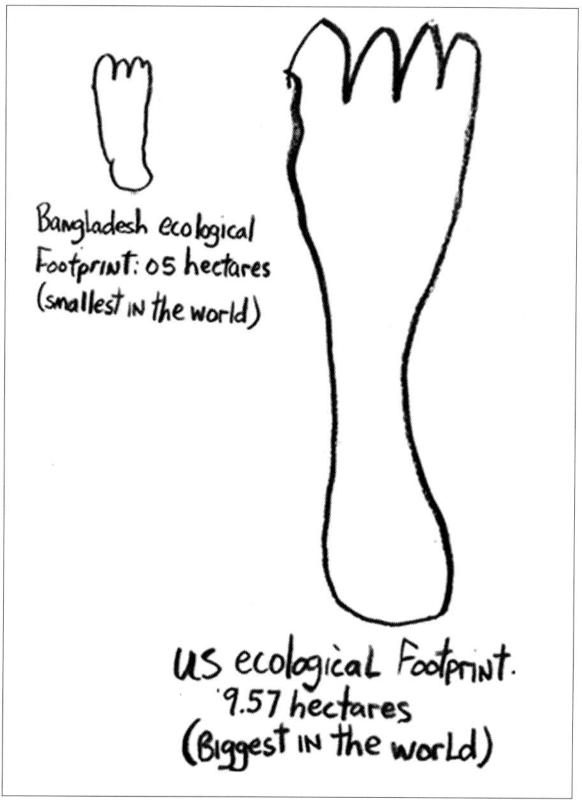

District official in Medak anointing a bullock cart during a festival celebrating local agricultural biodiversity.

Politics and Culture

The Cultural Politics of Food Sovereignty

The research and arguments presented in this publication articulate new perspectives on the meaning of agriculture and show the practical significance of defending local sources of food. They provide an important counterpoint to the idea that "food security" can only be ensured by the grace of the food industry and farms operating as industrial factories.

The concept of food security has been central to three decades of development discourse dealing with hunger and malnutrition. Initially, food security was seen primarily as a production problem, leading to a focus on international measures to break down barriers to the use of industrial food technologies and facilitate trade in food surpluses. The one-dimensional nature of this argument was later nuanced by recognition of the importance of constraints on access to food in the presence of abundance. Regions such as South Asia are largely self-sufficient in grain production but adequate calories and nutrition remain out of reach of people with too little money to purchase food on the market and too few other means through which to access food where and when they need it. This recognition stimulated a large number of studies on the multiple and inter-related factors affecting access to food in particular national and local settings, and increased awareness among academics and policy makers of the complexity of food distribution.[1] Responses to problems of food insecurity have since broadened to include support for initiatives that enhance access to assets such as land or employment needed to create food-secure livelihoods.

. . . these legally binding measures erode the sovereign foundation of the State in their citizens, and reinforce the relationship of the State with transnational corporations in the business of food and agriculture . . .

Despite the increasing sophistication of the food security debate, the relevance of the concept to the problem of hunger and malnutrition is being challenged by diverse social movements both in the North and the South. The term "food sovereignty" has emerged in the context of critical perspectives on the process of globalization and questions regarding the credibility of government and inter-governmental positions on the food problem.[2]

This shift is not simply a matter of rhetoric but rather is grounded in the perception that the capacity of the market and the State to address the food needs of the people has been seriously curtailed by multilateral trade regimes such as the World Trade Organization and numerous bilateral agreements among governments and private interests linking trade, aid and security arrangements. The concern is that these measures bind the State to the interests of transnational corporations in the business of food and agriculture. In a global food system dominated by a handful of transnational corporations, reliance on proprietary technologies and fickle commercial relations of international trade to feed the world is not only biased but also untenable: neither the State nor the corporations can reliably provide food to people who do not have the purchasing power to access food or farming technologies on the international market. The wisdom and authority of the State to enter into these deals, at least in the sphere of agriculture, is questioned.[3]

The distinguishing feature of the call for food sovereignty is its insistence on developing food production, circulation and consumption systems that are under the command and control of local communities, as against the increasing monopolization and control of the corporation and the State. This position differs not only from most arguments in the food security debate but also from the notion of "food rights" that informs the pronouncements of the World Food Summits.[4]

The affirmation of individual rights to food, while a useful demand in the political environments of the North, has not been incorporated wholeheartedly into the food sovereignty discourse because it does not directly address the right of communities to produce food and to retain command and control over local food systems. From a food sovereignty perspective, a focus on egocentric rights diverts attention from concrete economic and political relations such as corporate control over agricultural inputs and knowledge and economic policies that structure the global food system. It also runs the risk of reducing the issue of hunger and malnutrition to a humanitarian problem for the rich countries to solve, a prospect unacceptable to societies with long and rich agrarian histories.

The farming practices, cultural celebrations and political struggles of the Nayakrishi movement in Bangladesh and the Dalit Sanghams of the Deccan demonstrate concretely what food sovereignty can mean in a global and capitalist context. First, the practices are grounded in the development of ecological production systems that are localized rather than globalized. This strategy goes beyond the positive achievements of growing safe and 'organic' food. Organic agriculture can be another form of commercial monoculture, that is, growing one variety of a crop organically in a commercial agricultural system responding to the dictates of external markets and technologies.

The farming systems of the Deccan and Nayakrishi try to strike a balance between market and local determinations of food production by resisting capitalization. Agricultural inputs and related knowledge, including seed, soil amendments and agricultural implements, are generated locally and vigorously defended because they help ensure the capacity of the communities to initiate production cycles independently. "Sisters, keep seed in your hands" – is the call of the Nayakrishi farmer women to their sisters. This is not a political demand to establish private property in seed and genetic resources in favor of farming women. Rather, it evokes the tradition of a free flow of plant genetic resources within and between farming communities and the role of women as seed-keepers. This is a front line battle against private property claims in the sphere of nature and entails a new politics where the political realm is not independent from the production realm.

Second, the social relations underlying the production system are first of all ethical relations. Producing food ecologically and creating the conditions for the conservation of the gifts of nature is a way of relating to people, the external environment and other life forms that is ethical and peaceful by intent and design. It reproduces the material foundations of ethical community relations. In this context, food is not perceived strictly as a physical entity independent of the social relations that define where and how it is produced, distributed and consumed. Rather, food is a joy of life, produced and eaten not only to satisfy hunger but also to savor and share with others in the community.

By contrast, food stripped of its social character converges with the related notion of hunger as a strictly biological phenomenon, as though human beings were machines with big holes in their stomachs.

Consider, for example, the special role of the plant known as *Foshka begun* (Physalis heterophylla nees.) and its role in the life of children growing up in the rural areas of Bangladesh. When caring for a baby brother the older sister picks the soft green fruit of *Foshka begun* and presses it against her brother's forehead to make soft sounds to keep him amused. The relationship built by the plant between the brother and sister has been ritualized in the ceremony called *Bhai fota*. Sisters use the flower to make a stamp on the forehead of the brother on a particular day of Bengal's calendar. While the plant has no economic use it is revered for the role it plays in cementing and celebrating the relationship between people.

Once the notion of food as both a material and social entity is understood, the significance of uncultivated biodiversity in South Asia becomes strikingly visible. Thousands of farmers and food producing communities across South Asia still host a cornucopia of life on their farms and fields, both cultivated and uncultivated. The diversity nurtured by these women and men is an unbroken tradition, grounded in the continuity of oral culture and dynamic knowledge created in relation to food sources. These knowledge systems developed over thousands of years and yet are being discarded almost overnight.

The concept of food sovereignty provides a strategic focus for building new relationships between food producing communities and women`s movements, consumer`s movements, indigenous people`s movements, environmental movements and other communities of interest opposed to the patriarchal, homogenizing and destructive trends in the process of globalization. UBINIG and the Deccan Development Society, in collaboration with the South Asia Network on Food, Ecology and Culture (SANFEC), are actively engaged with farming communities in a wide range of initiatives to resist and redress the dangers presented by the rapid disarticulation of food systems. While these NGOs, as a matter of principle, do not claim to represent farmers or speak on behalf of the farming communities, the voices of farmers they work with are amplified through a form of cultural politics that is regional in scope and capable of articulating positive social policies.[5]

For example, the Deccan Development Society initiated a Biodiversity Festival in 1999 that quickly became a local initiative to chronicle and celebrate the culture of farming communities in a meaningful and politically engaging fashion. A procession or Mobile Biodiversity Festival is now organized annually by the Sanghams of the area, bringing the celebration directly to individual villages. Collections of local seeds are displayed inside carts drawn by oxen with painted horns and plaited braids. Village people wash entire streets, plaster with cow dung and draw ritual motifs to welcome them. When the bullocks walk these streets, people wash their feet, put holy marks on their foreheads and worship them. Singers, dancers and drummers join in, bringing their own stamp of novelty and creativity to the caravan. The songs they compose are sung in front of the church, the temple, the mosque and the *Sufi* shrine, embracing all faiths in a celebration of crops, seeds, livestock and agricultural implements, not particular gods. Evenings are marked by food festivals and screening of films on local agriculture made by local people, attracting huge crowds.[6] As the village festival ends, the carts are taken to the edge of the village where members of a neighbouring village pick them up with their own bullocks. This mobility ensures unprecedented participation from people, passing through more than 60 villages and engaging with over 100,000 farmers every year.

In the first years of the festival, people cried when they saw the many crops and varieties displayed in the carts and realized the seed wealth they had lost. Grandmothers opened their doors and asked that wealth return to their houses. The celebratory discourse of the caravan has since merged with substantive discussions of food and farming futures. In every village that the festival enters, people sit down to discuss the current situation and ways to address their concerns. The caravan always includes several *mandal* towns (representing a group of villages) and the district capital, making it possible to bring farmers and their concerns into direct contact with village, *mandal* and district level officials along with other members of the general population. Over time, the farmers have become very clear about what they will accept from governments and what they will reject. Top agricultural scientists, bureaucrats and politicians now come to the festivals to learn new perspectives on agrobiodiversity and gratefully honour farmers.

From these simple expressions of awareness among officials, small and larger shifts in policy perspective have emerged. In 2001, the festival led to the development of a Regional Action Plan for Agrobiodiversity that became a major part of the Government of India's National Biodiversity Strategy and Action Plan. The festivals also inspired negotiations leading to State Government support for an Alternative Public Distribution System replacing the rice and wheat imported

from other regions of India with millet, sorghum and pulses grown by farmers in the area. This alternative system reinforces local food culture and increases the market value of local grains. Over a period of six years the festivals have etched for themselves a permanent place in the cultural calendar of the community. They have also changed the perception that ecological agriculture in the Deccan is only an environmentalist's agenda, establishing beyond doubt the authenticity of the message and elevating farmer concerns to the national stage.

Biodiversity Festivals in combination with farmer exchanges are also part of the SANFEC regional strategy to accelerate the spread of ecological agriculture in South Asia and cultivate better understanding and relations between different communities. Over the last few years farmers from the flood plains of Bangladesh, the dry lands of India, the hills and plains of Nepal, the coastal and inland communities of Sri Lanka, Bangladesh and the Green Revolution areas of Pakistan have visited each other during and after these festivals, sharing experiences with ecological food production and related cultural practices. The knowledge and skill of farmers, as well as seed and genetic resources when appropriate, have flowed between communities divided by language, culture, religion and political systems.

Exchanges of farmers between Pakistan and Bangladesh, countries divided by bitter memories of their separation in 1971, and between southern India and the conflict zones of Sri Lanka, regions riddled with mutual mistrust, are vivid examples of grounded peace-building among farming communities. They are also a concrete expression of regional food sovereignty strategies that can enhance production practices and facilitate reciprocal sharing of strategic inputs.

The farmer exchanges organized through SANFEC occasionally go beyond the boundaries of South Asia by integrating farmers into international events and organizing

Sundaram Verma from Rajasthan, India visiting Chaina, Anwara and other farmers in Bangladesh.

farmer speaking tours. On one occasion a group of women and men farmers from five countries in South Asia traveled to western Canada where they spent 10 days in discussions and farm visits with Canadian organic farmers on issues related to farm subsidies, land stewardship, youth in farming, and organic farming methods. This was followed by a return visit to South Asia by Canadian farmers and participation in discussions with farmers, policy makers and scientists on the challenges and the struggles of Canadian farmers to rebuild ecological farms following decades of chemical-intensive farming. The process triggered new and comparative understanding of the existing commonalities among farming communities and the global dimensions of struggles to revitalize local food systems.[7]

The farmers participating in these exchanges, and the NGOs supporting them, recognize the need to go beyond the farming communities to build solidarity with other social groups working on issues of culture, human rights and social justice. Traditional folk singers, cultural activists, poets, painters, and creative writers have been strong allies of environmental and social movements and have played significant roles in disseminating the messages of the festivals strongly, widely and creatively. SANFEC has also worked closely with the Resistance Network, a regional network against trafficking in women and children, to raise awareness regarding the link between the destruction of food producing capacities, migration and trafficking. They played a crucial role in advocating for a convention to combat trafficking in women and children, later developed and signed by the governments of the South Asian Association for Regional Co-operation (SAARC). The two networks, along with other like-minded groups, jointly organize the SAARC People's Forum to introduce new issues onto the agenda of the annual SAARC Summits. In these events farmers play a direct role, contributing to a non-confrontational scenario of political advocacy difficult to achieve when non-governmental organizations act alone.

The food sovereignty strategies of farmers, SANFEC and its member organizations engage as well with the secular spiritual movements of the region that share the message of ecology and ethics that emerge from farming lifestyles. The various *Bhakti* traditions of Bengal, originating in different historical moments from Buddhism, Hinduism, Jainism and local religious sects, emphasize the profound role love and

Husband (Roshan, centre) and wife (Belu, left) are inseparable in the "Baul" community inspired by Lalon Fakir, a radical voice during British colonial rule.

caring play in organising the community and its material foundations in nature. Historically, the tradition emerged from rebellious grassroot movements challenging caste, class and gender hierarchies and fought to safeguard local knowledge as a form of secular spiritual experience. The wandering Baul musicians of Bangladesh embody this discourse through highly developed schools of poetry and music supported by a popular base of farmers, weavers, potters, fishers, jewelers, handicraft artists, and other marginalized groups. By reclaiming ecology and ethics in the context of spiritual knowledge, these traditions are able to highlight the significance of lifestyles as political and ethical behaviours. This establishes a direct link to the food sovereignty movements and their common purpose in opposition to violence, concern for social justice, and the practice of ecological ethics.

[1] *Various economists have highlighted the complexity of the food distribution problem, pointing in some cases to market failures that exclude communities without money from access to food and in others to the State policy of food distribution and the political nature of the State.*

[2] *The term "food sovereignty" forms part of the discourse of many social movements and non-governmental organizations including the Via Campesina (www.viacampesina.org), GRAIN (www.grain.org), Brazil's Landless Workers Movement (www.mstbrazil.org), Food First/Institute for Food and Development Policy (www.foodfirst.org) and the South Asia Network on Food, Ecology and Culture (www.sanfec.org), among others.*

[3] *The position of some non-governmental organizations promoting food sovereignty is that agriculture should remain outside of the WTO, leaving the State free to establish its own national food policies. Others argue that the WTO is the only legally binding vehicle capable of controlling the multiple bilateral trade and investment agreements opening countries to unfettered penetration by transnational corporations. For information on these concerns see: http://www.bilaterals.org).*

[4] *The 1996 Rome Declaration on World Food Security affirms "the right of everyone to have access to safe and nutritious food, consistent with the right to adequate food and the fundamental right of everyone to be free from hunger."*

[5] *SANFEC emerged initially as a vehicle for raising concerns regarding the agenda of the 1996 World Food Summit organized by the Food and Agriculture Organisation (FAO) of the United Nations. It subsequently evolved into a network of grassroots organizations and individuals committed to promoting peace and food sovereignty by strengthening the capacity of farming communities to contribute to policy dialogue and undertake grounded research on issues of concern to them. For more information see http://www.sanfec.org.*

[6] *Three farmer-made films accompany this publication. "Using Diversity: A Film by South Asian Farmers" is the story of biological wealth and cultural diversity as filmed by farmers in India, Nepal and Bangladesh. It brings us into farmers` fields, seed stores, household patios and common areas, helping us to see the knowledge behind people's actions and the need for policy change. "People's Agenda for Biodiversity" chronicles the journey of the Mobile Biodiversity Festivals over a period of several years. "South Asia – Canada Dialogue on the Future of Agriculture" depicts a remarkable encounter between farmers of South Asia and Canada. The message is loud and clear: food sovereignty means the conservation of diversity and ethical community relations.*

[7] *The statement "Common Ground: A Vision from the South Asia – Canada Dialogue on the Future of Agriculture" was presented by participating farmers at the Organic World Congress "Cultivating Communities" of the International Federation of Organic Agriculture Movements in Victoria, British Colombia, August 21-28th, 2002 and is posted to the SANFEC website.*

About the Authors

Farhad Mazhar (Bangladesh).

P.V. Satheesh (India).

Farhad Mazhar (Bangladesh) is a founding member of UBINIG (Policy Research for Development Alternative) and is currently its Managing Director. He has a degree in Economics from the New School for Social Research, New York and a degree in Pharmacy from the University of Dhaka, Bangladesh. He practiced pharmacy in New York for many years and contributed to establishing Gonoshaystha Pharmaceuticals, a firm in Bangladesh dedicated to enhancing national capacity to produce essential drugs for healthcare. He is best known for his poetry and regular columns in the major national newspapers of Bangladesh. In 1990, he facilitated the establishment of the Nayakrishi Andolon, a peasant's movement for biodiversity-based agriculture in Bangladesh. He is a founding animator of SANFEC (South Asia Network on Food, Ecology and Culture) and Chairman of the Editorial Board of *Chintaa*, a progressive fortnightly journal in Bangladesh.

P.V. Satheesh (India) is the co-founder and current Director of the Deccan Development Society. He is an accomplished filmmaker and development communication specialist, having worked for many years as a Senior Producer in Indian national television (Doordarshan) and a professor of communications at various Indian universities. He is on the Board of several national and international organizations, including the S.N. School of Communication (University of Hyderabad) and GRAIN, and is a founding animator of SANFEC (South Asia Network on Food, Ecology and Culture).

Daniel Buckles (Canada) is an Adjunct Research Professor in the Department of Sociology and Anthropology at Carleton University in Ottawa. He has published books on indigenous peoples, agriculture and conflict management and facilitated the development of a wide range of research and development projects in Asia, Latin America and Africa while working for the International Development Research Centre (IDRC). He was a Senior Scientist in the Economics Program with the International Maize and Wheat Improvement Centre (CIMMYT) based in Mexico and a co-founder of the Mexican NGO "Proyecto Sierra de Santa Marta". He is currently engaged in the development of concepts and tools for collaborative research and social action in the context of the Social Analysis Systems[2] initiative at Carleton University (www.sas-pm.com).

(right) Daniel Buckles (Canada)

(right) Farida Akhter (Bangladesh)

Farida Akhter (Bangladesh) is the Executive Director of UBINIG. She has an M.A. in Economics from Chittagong University, Bangladesh and started her career as a journalist before turning to social research on development, health, population issues, marine fisheries and agriculture. She is a leading member of many regional and international women's networks such as Feminist International Network of Resistance to Reproductive and Genetic Engineering (FINRRAGE), Asian Women's Human Rights Council (AWHRC) and United Women's Front in Bangladesh. She is the Editor of *Chintaa* and regular contributor to national dailies.

Acknowledgements

The studies that form the basis of this publication were collective explorations undertaken by farmers of Bangladesh and India with staff of UBINIG and the Deccan Development Society. The authors wish to acknowledge the fundamental contributions of these communities and the organizations we work with. We can only hope that the writing has been faithful to their inspiration and insight. While too many to name, the following individuals made contributions of great consequence. From Bangladesh: Jahanir Alam Jony, Rafiqul Huq Titu, Ruposhi Begum, Hazer Begum, Tamashi Begum, Rabeya Begum, Nahar, Rekha Begum, Nurjahan Begum Shakera Begum, Ansara Begum, Hasina Begum, Sabekun Nahar, Sakhina Begum, Pushpo Rani, Sameda Begum, Bayan Uddin, Rokeya Khanum Tulu, Shamsunnahar, Shahinoor Begum, Halima Parveen, Romela Akhter Rumi, Akkas Ali, Harun ur Rashid, Nazma Akhter, Dominque Caouette and Shima Das Shimu. From India: Begari Sammamma, Chinna Narsamma, Begari Laxmamma, Gangwar Manemma, Begari Susilamma, Begari Eeramma, Salome Yesudas and Suresh Reddy.

The design concept for the publication was developed by Daniel Buckles, who also edited the essays. Justin Wonnacott designed the publication and brought out the best in the photographs.

Anumu

Chinna Narsamma

Salome Yesudas

Anwara Begum

Yesu

Begari Sammamma (right) and Begari Laxmamma

Abdul Jabbar

H. Podinilame

Palash Barel

Rajeev Khedkar

Shahid Zia

Bhuwon Sthapit

Shree Ram Shrestha

Prabath Kumara

Ranjith de Silva

The animators of the South Asia Network on Food, Ecology and Culture (SANFEC) are (in alphabetical order): Farida Akhter (Bangladesh), Palash Barel (Bangladesh), Prabath Kumara (Sri Lanka), Farhad Mazhar (Bangladesh), H. Podinilame (Sri Lanka), Rajeev Khedkar (India), P.V. Satheesh (India), Shree Ram Shrestha (Nepal), R. de Silva (Sri Lanka), Bhuwon Sthapit (Nepal), and Shahid Zia (Pakistan).

Photo credits: Unless stated otherwise, photographs are by Daniel Buckles.

Page 3, upper right – Yesu; Page 5, upper right – Jabbar; Page 7, Peter and Lisa Kuhnlein; Page 9, lower right – Peter and Lisa Kuhnlein; Page 12, Jabbar; Page 15, lower right – Peter and Lisa Kuhnlein; Page 16, centre – Peter and Lisa Kuhnlein; left – Justin Wonnacott; Page 17 top & bottom – Peter and Lisa Kuhnlein; Page 18, upper left – Yesu; left – Justin Wonnacott; page 22 – Yesu; Page 25, greens strip – Peter and Lisa Kuhnlein; grains strip, – Yesu; page 26 right – Peter and Lisa Kuhnlein; page 27 – Justin Wonnacott; page 28 – Jabbar; page 34, – Peter and Lisa Kuhnlein; page 35 – Yesu; Page 36 – Yesu; page 37 – Peter and Lisa Kuhnlein; page 38 – Jabbar; page 39, middle – Jabbar; page 40, upper left and right, – Jabbar; page 44, – Jabbar; page 47, – Jabbar; page 49, – Yesu; page 51, upper right and lower center, – Peter and Lisa Kuhnlein; page 52, – Yesu; page 53, top – Jabbar; page 56, – Jabbar; page 58 left – Yesu; right – Peter and Lisa Kuhnlein; page 64, Sustainable Agriculture Action Group (SAAG), Pakistan; page 65, UBINIG; page 73, right, – Jabbar; page 74 (Yesudas & Yesu, – Peter and Lisa Kuhnlein; (Anumu) – Yesuin; page 75 (Prabath) – Yesu